T0296968

Cambridge Elementary Classics

THUCYDIDES

BOOK IV, CHAPTERS I—XLI.

(PYLUS AND SPHACTERIA)

THUCYDIDES

BOOK IV, CHAPTERS I–XLI.
(PYLUS AND SPHACTERIA)

Edited by

J. H. E. CREES

AND

J. C. WORDSWORTH

CAMBRIDGE
AT THE UNIVERSITY PRESS
1962

CAMBRIDGE UNIVERSITY PRESS
Cambridge, New York, Melbourne, Madrid, Cape Town, Singapore,
São Paulo, Delhi, Dubai, Tokyo

Cambridge University Press
The Edinburgh Building, Cambridge CB2 8RU, UK

Published in the United States of America by
Cambridge University Press, New York

www.cambridge.org
Information on this title: www.cambridge.org/9780521141178

© Cambridge University Press 1919

This publication is in copyright. Subject to statutory exception
and to the provisions of relevant collective licensing agreements,
no reproduction of any part may take place without the written
permission of Cambridge University Press.

First published 1919
Reprinted 1933, 1944, 1954, 1962
This digitally printed version 2010

A catalogue record for this publication is available from the British Library

ISBN 978-0-521-06634-1 Hardback
ISBN 978-0-521-14117-8 Paperback

Cambridge University Press has no responsibility for the persistence or
accuracy of URLs for external or third-party internet websites referred to in
this publication, and does not guarantee that any content on such websites is,
or will remain, accurate or appropriate.

PREFACE

THIS edition has been prepared for those who have not long been studying Greek and who have reached the stage of the "First School Examination." A vocabulary has therefore been added, in the hope that those using the book may thus make more rapid progress through it in the limited amount of time which too often remains for Greek when the demands of other subjects have been satisfied. The Introduction was written by Dr Crees, the Vocabulary was prepared by Mr Wordsworth, who also drafted the Notes after joint consultation.

J. H. E. CREES.
J. C. WORDSWORTH.

July 1919

CONTENTS

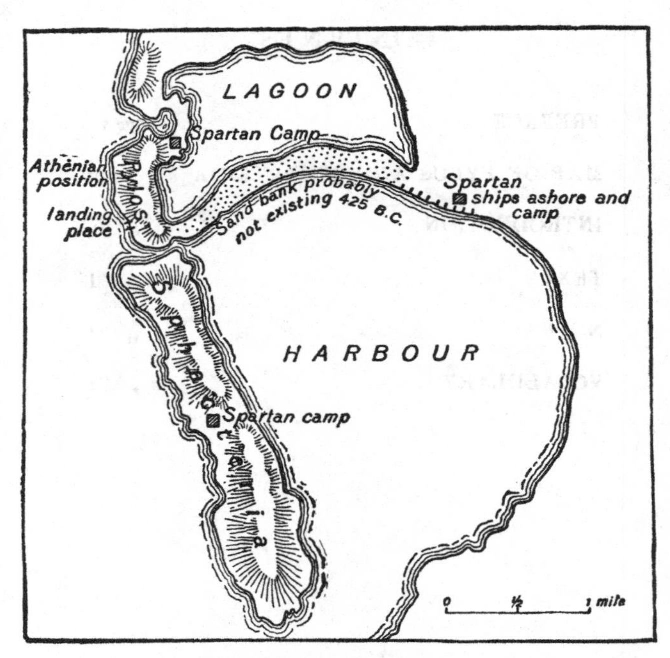

Map of Pylus and Sphacteria

INTRODUCTION

THUCYDIDES the son of Olorus was born probably about 471 B.C. He tells us little of himself but doubtless went through the same routine in his early years that all Athenians did. As an ephebus (aged 18–20) he would serve on garrison duty, and when he became a full Athenian citizen he must have served in some of the numerous campaigns in which Athens engaged. When the Peloponnesian War broke out (431 B.C.) he realised, he tells us, the greatness of the struggle, and prepared himself to write the history of it. We hear nothing definite of him till 421 B.C. when he was in command of an Athenian fleet stationed at Thasos[1]. Perhaps owing to his lack of vigour, Amphipolis on the Thracian mainland was taken by the Spartan Brasidas. The result of this reverse was banishment for Thucydides, who now had ample leisure for the composition of his history. Not until Athens had been overwhelmed and the War had ended (404 B.C.) was Thucydides able to return to Athens, but death came upon him before his task was finished. His history does not go past the year 411, and of the eight books which we have, two at least did not receive a final revision.

[1] Where Thucydides possessed gold mines which seem to have exercised an unhappy attraction on him.

As an historian Thucydides had been preceded by the charming and genial Herodotus (about 484–424 B.C.), whose delightful work was somewhat unjustly depreciated by Thucydides, according to the ways of literary men. And yet it is Thucydides perhaps rather than Herodotus who deserves the title "father of history." Thucydides' history would at any time have been a great work, but for its date it is in its conception a marvellous achievement, and the expression of a personality which compels respect. He put on one side graceful fictions, he disdained to tickle the ears of the groundlings with interesting but superficial narratives, and he resolved to reach to fundamentals. He conceived of history as a science and not as a fairytale, a record, as accurate and as full as possible, of things which have happened in the past and which, in all likelihood, will happen in similar fashion in the future. It is a permanent possession and not a rhetorical display entrancing and even thrilling and then forgotten. It would be difficult even in the present day to set forth better than Thucydides has done in a memorable chapter the function of the historian. Thucydides might have agreed with Goethe that "in this life we enact Hell." At any rate he strips life of all its embellishments, he lays bare human motives in all their stark crudity, he is the first of "real politicians." A city's existence depends upon the physical force which it can control, the struggle for existence is hard and bitter, and the weaker must go to the wall. If a city begins to walk in the ways of ambition it must show no mercy, it must forget good nature, it must not care too much about the good opinion of others. Even Athens the beautiful, which its citizens adore with the passion of a lover, is a tyrant city. Such ruthless doctrine, which at any rate truly represented the Greek world as Thucydides knew it, is not so forcibly expressed again until the time of Machiavelli, or later in the works of recent German philosopher-historians.

The rarest characteristic in historians (until we come to certain moderns who have no interests and no emotions) is the judicial mind. It is easy to be uninterested—and consequently uninteresting; it is terribly difficult to be disinterested. But the reserved and detached Thucydides has achieved even this. "Nothing extenuate and nought set down in malice" might have been his maxim. He could not, if he was a true Athenian, fail to be a partisan, and yet in the face of the most grievous injury which a man could suffer in party strife he effaced his partisanship and achieved a monumental impartiality. No man, one would think, could have despised and detested Cleon the tanner more than the haughty, reserved, and high-bred Thucydides. Yet, evident though the antipathy may be, the very fairness of the man represses all exaggeration, abuse, or bitterness, and the champions of Cleon must, and can, base their championship on the evidence of Thucydides. He was a just man.

Such a nature, austere, reserved, contemptuous of

superficiality, intolerant of exaggeration, found an apt expression in the Thucydidean style. It is unadorned, plain, cold, but in narrative remarkably clear. Thucydides "makes his readers beholders." He hated turgid rhapsody, and detested gush. All the great Greeks believed that the half is greater than the whole. They preferred restraint to excess, and left their readers' imagination some work to do. Thucydides is a master of economy. More than any other he pared down his sentiment. But the deep feeling is there, true, unforced and unmistakable, and any who read the story of the Athenian disaster in Sicily (Book VII) will at least understand the eulogy of Macaulay—no mean judge of greatness in an historian—"it is the *ne plus ultra* of human art."

Thucydides had few predecessors to help him in hammering out a style in Attic Prose. The graceful harmonies of Ionic were not for him; he had profounder things to say, things as yet unattempted. His ordinary narrative can be transparently clear, as simple as, or sometimes even simpler than, the unstudied ease of Xenophon[1]. But in the speeches which he inserts in his history (see in Book IV, chaps. 10, 17–20) we find him struggling to put into shape the thoughts and reflections to which different situations give rise, and making use of the rhetorical devices popularised by Gorgias the Sicilian, and

[1] The early chapters of Book IV are not the best specimens of Thucydides' easier style.

copied by Antiphon the Athenian (480–411 B.C.), of assonance, alliteration, and measured compass of phrase. It was a great task which he took upon himself, to give a more formal expression on the lines of the new rhetoric to the views of the statesmen and politicians of the time. It cannot be said that he by any means succeeded in giving his matter a perfect form. But "he went earnestly to work, yet with some glow of experimental daring; for it may be said that in antithesis and metaphor he found his only recreations[1]."

If Thucydides was fortunate in his subject, it is true also that Athens was fortunate to find such a narrator of her struggles. Those who care to trace analogies can find many a resemblance between the Athens which was the school of Greece, the city whose statesmen preached the gospel of brute force and declared that only ruthlessness gave security, and the Power of yesterday which claimed to dominate the world by Kultur and laughed at righteousness and compassion. Athens *was* a tyrant city, beautiful in her crown of violets, but pitiless and selfish. Yet no one has ever read the tale of the Peloponnesian War without feeling a real sympathy for that gracious city, queen of the Muses, whose downfall was so complete. That sympathy is the work of Thucydides.

[1] Lamb, *Clio Enthroned* (p. 162), a most interesting and valuable book written in a packed and Thucydidean style, which will however appeal rather to the mature scholar than the young student.

He has set forth Athens' faults, but he has given a noble expression to all those qualities which have made all succeeding generations, no less than her own citizens, her *lovers*.

The first forty chapters of Book IV are taken up mainly with the story of the events at Sphacteria. The duel between Athens and Sparta, the great military state of Greece, was something like a conflict between England and Germany, as it seemed to Bismarck, a contest between an elephant and a whale. Strong in her island-like defence of the Long Walls (which ran down from Athens to its port Piraeus) Athens could defy Sparta, if only she ignored the invasions of Attica and held fast. On the other hand, though she could raid the mainland of the Peloponnese in retaliation for the Spartan invasions she could do her enemy little real mischief. Thus the war begun in 432 had by 425 B.C. almost come to a standstill. Then, however, the Spartans, seeking to dislodge the Athenians from Pylus, a post in Messenia which the general Demosthenes had occupied, imprudently landed four hundred and twenty Spartans on the small adjacent island of Sphacteria. The men were completely cut off by the Athenian ships, much to the consternation of the Spartan government. But the Spartan soldiers had enormous prestige, and the Athenian commanders dallied. At length Cleon, the Athenian demagogue, who later was to drive Thucydides into exile, vigorously criticised the authorities and in a moment of impetuosity pledged himself to reduce Sphacteria within twenty days. The official party led by Nicias took him at his word and Cleon was compelled to go. He had probably taken the precaution to consult the general Demosthenes first[1]. He took no risks, overwhelmed his clumsy foes by mere numbers (10,000 to 420) and by skilful tactics, and easily fulfilled his pledge, mad though it seemed to Thucydides. It is an interesting glimpse of the party struggle at Athens, its pettiness and its bitterness, leading sometimes to blindness towards the common weal. "Cleon will not succeed, but if he fails, we shall be rid of him."

But he did not fail, and one hundred and twenty true-born Spartans were so precious to their state that Sparta was willing to buy peace on any reasonable terms. If Pericles had been alive he might have negotiated a lasting settlement, and the fourth century B.C. might have been as brilliant a period in Greek history as the fifth. Cleon however was no statesman. He bullied the Spartan envoys, and sent them packing. Ten years later Athens embarked upon the Syracusan expedition, failed, and slowly bled to death. But that is another and a longer story.

[1] It is quite likely that Demosthenes had enlisted the support of Cleon. Those who have studied modern municipal politics at any rate will incline to this view.

Some dates in the Peloponnesian War

(490 B.C. Marathon, 480 B.C. Thermopylae, 479 B.C. Salamis)

431 B.C. The war begins.
430 B.C. The plague at Athens.
429 B.C. Death of Pericles.
425 B.C. Capture of 120 Spartans at Sphacteria.
415–413 B.C. Athenian expedition to Syracuse, and failure.
412 B.C. Revolt of the subject allies.
404 B.C. Athens surrenders to Sparta.

ΘΟΥΚΥΔΙΔΟΥ ΞΥΓΓΡΑΦΗΣ Δ

I. Events in Sicily. Messana revolts from Athens. The Locrians attack Rhegium.

Τοῦ δ' ἐπιγιγνομένου θέρους περὶ σίτου ἐκβολὴν Συρακοσίων δέκα νῆες πλεύσασαι καὶ Λοκρίδες ἴσαι Μεσσήνην τὴν ἐν Σικελίᾳ κατέλαβον, αὐτῶν ἐπαγαγομένων, καὶ ἀπέστη Μεσσήνη Ἀθηναίων. ἔπραξαν 2 δὲ τοῦτο μάλιστα οἱ μὲν Συρακόσιοι ὁρῶντες προσβολὴν ἔχον τὸ χωρίον τῆς Σικελίας καὶ φοβούμενοι τοὺς Ἀθηναίους μὴ ἐξ αὐτοῦ ὁρμώμενοί ποτε σφίσι μείζονι παρασκευῇ ἐπέλθωσιν, οἱ δὲ Λοκροὶ κατὰ ἔχθος τὸ Ῥηγίνων, βουλόμενοι ἀμφοτέρωθεν αὐτοὺς καταπολεμεῖν. καὶ ἐσεβεβλήκεσαν ἅμα ἐς τὴν Ῥη- 3 γίνων οἱ Λοκροὶ πανστρατιᾷ, ἵνα μὴ ἐπιβοηθῶσι τοῖς Μεσσηνίοις, ἅμα δὲ καὶ ξυνεπαγόντων Ῥηγίνων φυγάδων, οἳ ἦσαν παρ' αὐτοῖς· τὸ γὰρ Ῥήγιον ἐπὶ πολὺν χρόνον ἐστασίαζε καὶ ἀδύνατα ἦν ἐν τῷ παρόντι τοὺς Λοκροὺς ἀμύνεσθαι, ᾗ καὶ μᾶλλον ἐπετίθεντο. δῃώ- 4 σαντες δὲ οἱ μὲν Λοκροὶ τῷ πεζῷ ἀπεχώρησαν, αἱ δὲ νῆες Μεσσήνην ἐφρούρουν· καὶ ἄλλαι [αἱ] πληρούμεναι ἔμελλον αὐτόσε ἐγκαθορμισάμεναι τὸν πόλεμον ἐντεῦθεν ποιήσεσθαι.

II. Invasion of Attica by the Peloponnesians. Athenian and Spartan fleets sail for Corcyra.

Ὑπὸ δὲ τοὺς αὐτοὺς χρόνους τοῦ ἦρος, πρὶν τὸν σῖτον ἐν ἀκμῇ εἶναι, Πελοποννήσιοι καὶ οἱ ξύμμαχοι

ἐσέβαλον ἐς τὴν Ἀττικήν (ἡγεῖτο δὲ Ἆγις ὁ Ἀρχιδάμου, Λακεδαιμονίων βασιλεύς), καὶ ἐγκαθεζόμενοι
2 ἐδῄουν τὴν γῆν. Ἀθηναῖοι δὲ τάς τε τεσσαράκοντα ναῦς ἐς Σικελίαν ἀπέστειλαν, ὥσπερ παρεσκευάζοντο, καὶ στρατηγοὺς τοὺς ὑπολοίπους Εὐρυμέδοντα καὶ Σοφοκλέα· Πυθόδωρος γὰρ ὁ τρίτος αὐτῶν ἤδη
3 προαφῖκτο ἐς Σικελίαν. εἶπον δὲ τούτοις και Κερκυραίων ἅμα παραπλέοντας τῶν ἐν τῇ πόλει ἐπιμεληθῆναι, οἳ ἐλῃστεύοντο ὑπὸ τῶν ἐν τῷ ὄρει φυγάδων· καὶ Πελοποννησίων αὐτόσε νῆες ἑξήκοντα παρεπεπλεύκεσαν τοῖς ἐν τῷ ὄρει τιμωροί, καὶ λιμοῦ ὄντος μεγάλου ἐν τῇ πόλει νομίζοντες κατασχήσειν ῥᾳδίως
4 τὰ πράγματα. Δημοσθένει δὲ ὄντι ἰδιώτῃ μετὰ τὴν ἀναχώρησιν τὴν ἐξ Ἀκαρνανίας αὐτῷ δεηθέντι εἶπον χρῆσθαι ταῖς ναυσὶ ταύταις, ἢν βούληται, περὶ τὴν Πελοπόννησον.

III. Demosthenes advises the seizure and fortification of Pylus.

Καὶ ὡς ἐγένοντο πλέοντες κατὰ τὴν Λακωνικὴν καὶ ἐπυνθάνοντο ὅτι αἱ νῆες ἐν Κερκύρᾳ ἤδη εἰσὶ τῶν Πελοποννησίων, ὁ μὲν Εὐρυμέδων καὶ Σοφοκλῆς ἠπείγοντο ἐς τὴν Κέρκυραν, ὁ δὲ Δημοσθένης ἐς τὴν Πύλον πρῶτον ἐκέλευε σχόντας αὐτοὺς καὶ πράξαντας ἃ δεῖ τὸν πλοῦν ποιεῖσθαι· ἀντιλεγόντων δὲ κατὰ τύχην χειμὼν ἐπιγενόμενος κατήνεγκε τὰς ναῦς ἐς τὴν Πύ-
2 λον. καὶ ὁ Δημοσθένης εὐθὺς ἠξίου τειχίζεσθαι τὸ χωρίον (ἐπὶ τοῦτο γὰρ ξυνεκπλεῦσαι), καὶ ἀπέφαινε πολλὴν εὐπορίαν ξύλων τε καὶ λίθων καὶ φύσει καρτερὸν ὂν καὶ ἐρῆμον αὐτό τε καὶ ἐπὶ πολὺ τῆς χώρας· ἀπέχει γὰρ σταδίους μάλιστα ἡ Πύλος τῆς Σπάρτης τετρακοσίους, καὶ ἔστιν ἐν τῇ Μεσσηνίᾳ ποτὲ οὔσῃ

γῇ, καλοῦσι δὲ αὐτὴν οἱ Λακεδαιμόνιοι Κορυφάσιον. οἱ δὲ πολλὰς ἔφασαν εἶναι ἄκρας ἐρήμους τῆς Πελο- 3 ποννήσου, ἣν βούληται καταλαμβάνων τὴν πόλιν δαπανᾶν. τῷ δὲ διάφορόν τι ἐδόκει εἶναι τοῦτο τὸ χωρίον ἑτέρου μᾶλλον, λιμένος τε προσόντος, καὶ τοὺς Μεσσηνίους οἰκείους ὄντας αὐτῷ τὸ ἀρχαῖον καὶ ὁμοφώνους τοῖς Λακεδαιμονίοις πλεῖστ' ἂν βλάπτειν ἐξ αὐτοῦ ὁρμωμένους καὶ βεβαίους ἅμα τοῦ χωρίου φύλακας ἔσεσθαι.

IV. Fortification of Pylus.

Ὡς δὲ οὐκ ἔπειθεν οὔτε τοὺς στρατηγοὺς οὔτε τοὺς στρατιώτας, ὕστερον καὶ τοῖς ταξιάρχοις κοινώσας, ἡσύχαζεν ὑπὸ ἀπλοίας, μέχρι αὐτοῖς τοῖς στρατιώταις σχολάζουσιν ὁρμὴ ἐνέπεσε περιστᾶσιν ἐκτειχίσαι τὸ χωρίον. καὶ ἐγχειρήσαντες εἰργάζοντο, 2 σιδήρια μὲν λιθουργὰ οὐκ ἔχοντες, λογάδην δὲ φέροντες λίθους, καὶ ξυνετίθεσαν ὡς ἕκαστόν τι ξυμβαίνοι· καὶ τὸν πηλόν, εἴ που δέοι χρῆσθαι, ἀγγείων ἀπορίᾳ ἐπὶ τοῦ νώτου ἔφερον ἐγκεκυφότες τε, ὡς μάλιστα μέλλοι ἐπιμένειν, καὶ τὼ χεῖρε ἐς τοὐπίσω ξυμπλέκοντες, ὅπως μὴ ἀποπίπτοι. παντί τε τρόπῳ 3 ἠπείγοντο φθῆναι τοὺς Λακεδαιμονίους τὰ ἐπιμαχώτατα ἐξεργασάμενοι πρὶν ἐπιβοηθῆσαι· τὸ γὰρ πλέον τοῦ χωρίου αὐτὸ καρτερὸν ὑπῆρχε καὶ οὐδὲν ἔδει τείχους.

V. Indifference at Sparta. Demosthenes is left with five ships at Pylus.

Οἱ δὲ ἑορτήν τινα ἔτυχον ἄγοντες, καὶ ἅμα πυνθανόμενοι ἐν ὀλιγωρίᾳ ἐποιοῦντο, ὡς, ὅταν ἐξέλθωσιν, ἢ οὐχ ὑπομενοῦντας σφᾶς ἢ ῥᾳδίως

ληψόμενοι βίᾳ· καί τι καὶ αὐτοὺς ὁ στρατὸς ἔτι ἐν
2 ταῖς Ἀθήναις ὢν ἐπέσχε. τειχίσαντες δὲ οἱ Ἀθηναῖοι τοῦ χωρίου τὰ πρὸς ἤπειρον καὶ ἃ μάλιστα ἔδει ἐν ἡμέραις ἓξ τὸν μὲν Δημοσθένη μετὰ νεῶν πέντε αὐτοῦ φύλακα καταλείπουσι, ταῖς δὲ πλειοσὶ ναυσὶ τὸν ἐς τὴν Κέρκυραν πλοῦν καὶ Σικελίαν ἠπείγοντο.

VI. Return of the Spartan army from Attica.

Οἱ δ' ἐν τῇ Ἀττικῇ ὄντες Πελοποννήσιοι ὡς ἐπύθοντο τῆς Πύλου κατειλημμένης, ἀνεχώρουν κατὰ τάχος ἐπ' οἴκου, νομίζοντες μὲν οἱ Λακεδαιμόνιοι καὶ Ἆγις ὁ βασιλεὺς οἰκεῖον σφίσι τὸ περὶ τὴν Πύλον· ἅμα δὲ πρῴ ἐσβαλόντες καὶ τοῦ σίτου ἔτι χλωροῦ ὄντος ἐσπάνιζον τροφῆς τοῖς πολλοῖς, χειμών τε ἐπιγενόμενος μείζων παρὰ τὴν καθεστηκυῖαν ὥραν
2 ἐπίεσε τὸ στράτευμα. ὥστε πολλαχόθεν ξυνέβη ἀναχωρῆσαί τε θᾶσσον αὐτοὺς καὶ βραχυτάτην γενέσθαι τὴν ἐσβολὴν ταύτην· ἡμέρας γὰρ πεντεκαίδεκα ἔμειναν ἐν τῇ Ἀττικῇ.

VII. Athenian failure at Eion.

Κατὰ δὲ τὸν αὐτὸν χρόνον Σιμωνίδης Ἀθηναίων στρατηγὸς Ἠιόνα τὴν ἐπὶ Θρᾴκης Μενδαίων ἀποικίαν, πολεμίαν δὲ οὖσαν, ξυλλέξας Ἀθηναίους τε ὀλίγους ἐκ τῶν φρουρίων καὶ τῶν ἐκείνῃ ξυμμάχων πλῆθος προδιδομένην κατέλαβε. καὶ παραχρῆμα ἐπιβοηθησάντων Χαλκιδέων καὶ Βοττιαίων ἐξεκρούσθη τε καὶ ἀπέβαλε πολλοὺς τῶν στρατιωτῶν.

VIII. The Spartans march against Pylus, and send for their fleet from Corcyra. Demosthenes warns the Athenian fleet. Description of Sphacteria. Occupation of the island by the Spartans.

Ἀναχωρησάντων δὲ τῶν ἐκ τῆς Ἀττικῆς Πελοποννησίων, οἱ Σπαρτιᾶται αὐτοὶ μὲν καὶ οἱ ἐγγύτατα τῶν περιοίκων εὐθὺς ἐβοήθουν ἐπὶ τὴν Πύλον, τῶν δὲ ἄλλων Λακεδαιμονίων βραδυτέρα ἐγίγνετο ἡ ἔφοδος, ἄρτι ἀφιγμένων ἀφ' ἑτέρας στρατείας. περιήγγελλον δὲ καὶ κατὰ τὴν Πελοπόννησον βοηθεῖν ὅτι τάχιστα ἐπὶ Πύλον, καὶ ἐπὶ τὰς ἐν τῇ Κερκύρᾳ ναῦς σφῶν τὰς ἑξήκοντα ἔπεμψαν, αἳ ὑπερενεχθεῖσαι τὸν Λευκαδίων ἰσθμὸν καὶ λαθοῦσαι τὰς ἐν Ζακύνθῳ Ἀττικὰς ναῦς ἀφικνοῦνται ἐπὶ Πύλον· παρῆν δὲ ἤδη καὶ ὁ πεζὸς στρατός. Δημοσθένης δὲ προσπλεόντων ἔτι τῶν Πελοποννησίων ὑπεκπέμπει φθάσας δύο ναῦς ἀγγεῖλαι Εὐρυμέδοντι καὶ τοῖς ἐν ταῖς ναυσὶν ἐν Ζακύνθῳ Ἀθηναίοις παρεῖναι ὡς τοῦ χωρίου κινδυνεύοντος. καὶ αἱ μὲν νῆες κατὰ τάχος ἔπλεον κατὰ τὰ ἐπεσταλμένα ὑπὸ Δημοσθένους· οἱ δὲ Λακεδαιμόνιοι παρεσκευάζοντο ὡς τῷ τειχίσματι προσβαλοῦντες κατά τε γῆν καὶ κατὰ θάλασσαν, ἐλπίζοντες ῥᾳδίως αἱρήσειν οἰκοδόμημα διὰ ταχέων εἰργασμένον καὶ ἀνθρώπων ὀλίγων ἐνόντων. προσδεχόμενοι δὲ καὶ τὴν ἀπὸ τῆς Ζακύνθου τῶν Ἀττικῶν νεῶν βοήθειαν ἐν νῷ εἶχον, ἢν ἄρα μὴ πρότερον ἕλωσι, καὶ τοὺς ἔσπλους τοῦ λιμένος ἐμφάρξαι, ὅπως μὴ ᾖ τοῖς Ἀθηναίοις ἐφορμίσασθαι ἐς αὐτόν. ἡ γὰρ νῆσος ἡ Σφακτηρία καλουμένη τόν τε λιμένα, παρατείνουσα καὶ ἐγγὺς ἐπικειμένη, ἐχυρὸν ποιεῖ καὶ τοὺς ἔσπλους στενούς, τῇ μὲν δυοῖν νεοῖν διάπλουν κατὰ

(margin section numbers: 2, 3, 4, 5, 6)

τὸ τείχισμα τῶν Ἀθηναίων καὶ τὴν Πύλον, τῇ δὲ πρὸς τὴν ἄλλην ἤπειρον ὀκτὼ ἢ ἐννέα· ὑλώδης τε καὶ ἀτριβὴς πᾶσα ὑπ' ἐρημίας ἦν, καὶ μέγεθος περὶ 7 πέντε καὶ δέκα σταδίους μάλιστα. τοὺς μὲν οὖν ἔσπλους ταῖς ναυσὶν ἀντιπρῴροις βύζην κλῄσειν ἔμελλον· τὴν δὲ νῆσον ταύτην φοβούμενοι μὴ ἐξ αὐτῆς τὸν πόλεμον σφίσι ποιῶνται, ὁπλίτας διεβίβασαν εἰς αὐτὴν καὶ παρὰ τὴν ἤπειρον ἄλλους 8 ἔταξαν. οὕτω γὰρ τοῖς Ἀθηναίοις τήν τε νῆσον πολεμίαν ἔσεσθαι τήν τε ἤπειρον, ἀπόβασιν οὐκ ἔχουσαν (τὰ γὰρ αὐτῆς τῆς Πύλου ἔξω τοῦ ἔσπλου πρὸς τὸ πέλαγος ἀλίμενα ὄντα οὐχ ἕξειν ὅθεν ὁρμώμενοι ὠφελήσουσι τοὺς αὑτῶν), σφεῖς δὲ ἄνευ τε ναυμαχίας καὶ κινδύνου ἐκπολιορκήσειν τὸ χωρίον κατὰ τὸ εἰκός, σίτου τε οὐκ ἐνόντος καὶ δι' ὀλίγης 9 παρασκευῆς κατειλημμένον. ὡς δ' ἐδόκει αὐτοῖς ταῦτα, καὶ διεβίβαζον ἐς τὴν νῆσον τοὺς ὁπλίτας ἀποκληρώσαντες ἀπὸ πάντων τῶν λόχων. καὶ διέβησαν μὲν καὶ ἄλλοι πρότερον κατὰ διαδοχήν, οἱ δὲ τελευταῖοι καὶ ἐγκαταληφθέντες εἴκοσι καὶ τετρακόσιοι ἦσαν καὶ Εἵλωτες οἱ περὶ αὐτούς· ἦρχε δὲ αὐτῶν Ἐπιτάδας ὁ Μολόβρου.

IX. Arrangements of Demosthenes to meet the attack.

Δημοσθένης δὲ ὁρῶν τοὺς Λακεδαιμονίους μέλλοντας προσβάλλειν ναυσί τε ἅμα καὶ πεζῷ, παρεσκευάζετο καὶ αὐτός, καὶ τὰς τριήρεις αἳ περιῆσαν αὐτῷ ἀπὸ τῶν καταλειφθεισῶν ἀνασπάσας ὑπὸ τὸ τείχισμα προσεσταύρωσε, καὶ τοὺς ναύτας ἐξ αὐτῶν ὥπλισεν ἀσπίσι τε φαύλαις καὶ οἰσυΐναις ταῖς πολλαῖς· οὐ γὰρ ἦν ὅπλα ἐν χωρίῳ ἐρήμῳ πορίσασθαι, ἀλλὰ καὶ ταῦτα ἐκ λῃστρικῆς Μεσσηνίων τριακοντόρου καὶ κέλητος ἔλαβον, οἳ ἔτυχον παραγενόμενοι. ὁπλῖταί τε τῶν Μεσσηνίων τούτων ὡς τεσσαράκοντα ἐγένοντο, οἷς ἐχρῆτο μετὰ τῶν ἄλλων. τοὺς μὲν οὖν πολλοὺς τῶν τε ἀόπλων καὶ ὡπλισμένων 2 ἐπὶ τὰ τετειχισμένα μάλιστα καὶ ἐχυρὰ τοῦ χωρίου πρὸς τὴν ἤπειρον ἔταξε, προειπὼν ἀμύνασθαι τὸν πεζόν, ἢν προσβάλῃ· αὐτὸς δὲ ἀπολεξάμενος ἐκ πάντων ἑξήκοντα ὁπλίτας καὶ τοξότας ὀλίγους ἐχώρει ἔξω τοῦ τείχους ἐπὶ τὴν θάλασσαν, ᾗ μάλιστα ἐκείνους προσεδέχετο πειράσειν ἀποβαίνειν ἐς χωρία μὲν χαλεπὰ καὶ πετρώδη πρὸς τὸ πέλαγος τετραμμένα, σφίσι δὲ τοῦ τείχους ταύτῃ ἀσθενεστάτου ὄντος ἐσβιάσασθαι αὐτοὺς ἡγεῖτο προθυμήσεσθαι· οὔτε γὰρ αὐτοὶ ἐλπίζοντές ποτε ναυσὶ κρατηθήσεσθαι 3 οὐκ ἰσχυρὸν ἐτείχιζον, ἐκείνοις τε βιαζομένοις τὴν ἀπόβασιν ἁλώσιμον τὸ χωρίον γίγνεσθαι. κατὰ 4 τοῦτο οὖν πρὸς αὐτὴν τὴν θάλασσαν χωρήσας ἔταξε τοὺς ὁπλίτας ὡς εἴρξων ἢν δύνηται, καὶ παρεκελεύσατο τοιάδε.

X. Speech of Demosthenes.

"Ἄνδρες οἱ ξυναράμενοι τοῦδε τοῦ κινδύνου, μηδεὶς ὑμῶν ἐν τῇ τοιᾷδε ἀνάγκῃ ξυνετὸς βουλέσθω δοκεῖν εἶναι, ἐκλογιζόμενος ἅπαν τὸ περιεστὸς ἡμᾶς δεινόν, μᾶλλον δ' ἀπερισκέπτως εὔελπις ὁμόσε χωρῆσαι τοῖς ἐναντίοις καὶ ἐκ τούτων ἂν περιγενόμενος. ὅσα γὰρ ἐς ἀνάγκην ἀφῖκται ὥσπερ τάδε, λογισμὸν ἥκιστα ἐνδεχόμενα κινδύνου τοῦ ταχίστου προσδεῖται. ἐγὼ δὲ καὶ τὰ πλείω ὁρῶ πρὸς ἡμῶν 2 ὄντα, ἢν ἐθέλωμέν τε μεῖναι καὶ μὴ τῷ πλήθει αὐτῶν

καταπλαγέντες τὰ ὑπάρχοντα ἡμῖν κρείσσω κατα-
3 προδοῦναι. τοῦ τε γὰρ χωρίου τὸ δυσέμβατον ἡμέτερον νομίζω, < ὃ > μενόντων ἡμῶν ξύμμαχον γίγνεται, ὑποχωρήσασι δὲ καίπερ χαλεπὸν ὂν εὔπορον ἔσται μηδενὸς κωλύοντος· καὶ τὸν πολέμιον δεινότερον ἕξομεν μὴ ῥᾳδίως αὐτῷ πάλιν οὔσης τῆς ἀναχωρήσεως, ἢν καὶ ὑφ' ἡμῶν βιάζηται (ἐπὶ γὰρ ταῖς ναυσὶ ῥᾷστοί
4 εἰσιν ἀμύνεσθαι, ἀποβάντες δὲ ἐν τῷ ἴσῳ ἤδη), τό τε πλῆθος αὐτῶν οὐκ ἄγαν δεῖ φοβεῖσθαι· κατ' ὀλίγον γὰρ μαχεῖται καίπερ πολὺ ὂν ἀπορίᾳ τῆς προσορμίσεως, καὶ οὐκ ἐν γῇ στρατός ἐστιν ἐκ τοῦ ὁμοίου μείζων, ἀλλ' ἀπὸ νεῶν, αἷς πολλὰ τὰ καίρια δεῖ ἐν
5 τῇ θαλάσσῃ ξυμβῆναι. ὥστε τὰς τούτων ἀπορίας ἀντιπάλους ἡγοῦμαι τῷ ἡμετέρῳ πλήθει, καὶ ἅμα ἀξιῶ ὑμᾶς Ἀθηναίους ὄντας καὶ ἐπισταμένους ἐμπειρίᾳ τὴν ναυτικὴν ἐπ' ἄλλους ἀπόβασιν, ὅτι εἴ τις ὑπομένοι καὶ μὴ φόβῳ ῥοθίου καὶ νεῶν δεινότητος κατάπλου ὑποχωροίη, οὐκ ἄν ποτε βιάζοιτο, καὶ αὐτοὺς νῦν μεῖναί τε καὶ ἀμυνομένους παρ' αὐτὴν τὴν ῥαχίαν σῴζειν ὑμᾶς τε αὐτοὺς καὶ τὸ χωρίον."

XI. Attack on Pylus by sea and land. Brasidas distinguishes himself on the Spartan side.

Τοσαῦτα τοῦ Δημοσθένους παρακελευσαμένου οἱ Ἀθηναῖοι ἐθάρσησάν τε μᾶλλον καὶ ἐπικαταβάντες
2 ἐτάξαντο παρ' αὐτὴν τὴν θάλασσαν. οἱ δὲ Λακεδαιμόνιοι ἄραντες τῷ τε κατὰ γῆν στρατῷ προσέβαλλον τῷ τειχίσματι καὶ ταῖς ναυσὶν ἅμα, οὔσαις τεσσαράκοντα καὶ τρισί· ναύαρχος δὲ αὐτῶν ἐπέπλει Θρασυμηδίδας ὁ Κρατησικλέους Σπαρτιάτης. προσ-
3 έβαλλε δὲ ᾗπερ ὁ Δημοσθένης προσεδέχετο. καὶ

οἱ μὲν Ἀθηναῖοι ἀμφοτέρωθεν, ἔκ τε γῆς καὶ ἐκ θαλάσσης, ἠμύνοντο· οἱ δὲ κατ' ὀλίγας ναῦς διελόμενοι, διότι οὐκ ἦν πλείοσι προσσχεῖν, καὶ ἀναπαύοντες ἐν τῷ μέρει τοὺς ἐπίπλους ἐποιοῦντο, προθυμίᾳ τε πάσῃ χρώμενοι καὶ παρακελευσμῷ, εἴ πως ὠσάμενοι ἕλοιεν τὸ τείχισμα. πάντων δὲ φανερώτατος Βρασίδας ἐγένετο. τριηραρχῶν γὰρ καὶ ὁρῶν τοῦ χωρίου χαλεποῦ ὄντος τοὺς τριηράρχους καὶ κυβερνήτας, εἴ πῃ και δοκοίη δυνατὸν εἶναι σχεῖν, ἀποκνοῦντας καὶ φυλασσομένους τῶν νεῶν μὴ ξυντρίψωσιν, ἐβόα λέγων ὡς οὐκ εἰκὸς εἴη ξύλων φειδομένους τοὺς πολεμίους ἐν τῇ χώρᾳ περιιδεῖν τεῖχος πεποιημένους, ἀλλὰ τάς τε σφετέρας ναῦς βιαζομένους τὴν ἀπόβασιν καταγνύναι ἐκέλευε καὶ τοὺς ξυμμάχους μὴ ἀποκνῆσαι ἀντὶ μεγάλων εὐεργεσιῶν τὰς ναῦς τοῖς Λακεδαιμονίοις ἐν τῷ παρόντι ἐπιδοῦναι, ὀκείλαντας δὲ καὶ παντὶ τρόπῳ ἀποβάντας τῶν τε ἀνδρῶν καὶ τοῦ χωρίου κρατῆσαι. 4

XII. Failure of the attack.

Καὶ ὁ μὲν τούς τε ἄλλους τοιαῦτα ἐπέσπερχε καὶ τὸν ἑαυτοῦ κυβερνήτην ἀναγκάσας ὀκεῖλαι τὴν ναῦν ἐχώρει ἐπὶ τὴν ἀποβάθραν· καὶ πειρώμενος ἀποβαίνειν ἀνεκόπη ὑπὸ τῶν Ἀθηναίων, καὶ τραυματισθεὶς πολλὰ ἐλιποψύχησέ τε, καὶ πεσόντος αὐτοῦ ἐς τὴν παρεξειρεσίαν ἡ ἀσπὶς περιερρύη ἐς τὴν θάλασσαν, καὶ ἐξενεχθείσης αὐτῆς ἐς τὴν γῆν οἱ Ἀθηναῖοι ἀνελόμενοι ὕστερον πρὸς τὸ τροπαῖον ἐχρήσαντο ὃ ἔστησαν τῆς προσβολῆς ταύτης. οἱ δ' ἄλλοι προυθυμοῦντο μὲν ἀδύνατοι δ' ἦσαν ἀποβῆναι τῶν τε χωρίων χαλεπότητι καὶ τῶν Ἀθηναίων 2

3 μενόντων καὶ οὐδὲν ὑποχωρούντων. ἐς τοῦτό τε περιέστη ἡ τύχη ὥστε Ἀθηναίους μὲν ἐκ γῆς τε καὶ ταύτης Λακωνικῆς ἀμύνεσθαι ἐκείνους ἐπιπλέοντας, Λακεδαιμονίους δὲ ἐκ νεῶν τε καὶ ἐς τὴν ἑαυτῶν πολεμίαν οὖσαν ἐπ' Ἀθηναίους ἀποβαίνειν. ἐπὶ πολὺ γὰρ † ἐποίει τῆς δόξης † ἐν τῷ τότε τοῖς μὲν ἠπειρώταις μάλιστα εἶναι καὶ τὰ πεζὰ κρατίστοις, τοῖς δὲ θαλασσίοις τε καὶ ταῖς ναυσὶ πλεῖστον προὔχειν.

XIII. Arrival of the Athenian fleet.

Ταύτην μὲν οὖν τὴν ἡμέραν καὶ τῆς ὑστεραίας μέρος τι προσβολὰς ποιησάμενοι ἐπέπαυντο· καὶ τῇ τρίτῃ ἐπὶ ξύλα ἐς μηχανὰς παρέπεμψαν τῶν νεῶν τινὰς ἐς Ἀσίνην, ἐλπίζοντες τὸ κατὰ τὸν λιμένα τεῖχος ὕψος μὲν ἔχειν, ἀποβάσεως δὲ μάλιστα οὔσης

2 ἑλεῖν μηχαναῖς. ἐν τούτῳ δὲ αἱ ἐκ τῆς Ζακύνθου νῆες τῶν Ἀθηναίων παραγίγνονται τεσσαράκοντα· προσεβοήθησαν γὰρ τῶν τε φρουρίδων τινες αὐτοῖς

3 τῶν ἐκ Ναυπάκτου καὶ Χῖαι τέσσαρες. ὡς δε εἶδον τήν τε ἤπειρον ὁπλιτῶν περίπλεων τήν τε νῆσον, ἔν τε τῷ λιμένι οὔσας τὰς ναῦς καὶ οὐκ ἐκπλεούσας, ἀπορήσαντες ὅπῃ καθορμίσωνται, τότε μὲν ἐς Πρωτὴν τὴν νῆσον, ἣ οὐ πολὺ ἀπέχει ἐρῆμος οὖσα, ἔπλευσαν καὶ ηὐλίσαντο, τῇ δ' ὑστεραίᾳ παρασκευασάμενοι ὡς ἐπὶ ναυμαχίαν ἀνήγοντο, ἢν μὲν ἀντεκπλεῖν ἐθέλωσι σφίσιν ἐς τὴν εὐρυχωρίαν, εἰ δὲ μή, ὡς αὐτοὶ

4 ἐπεσπλευσούμενοι. καὶ οἱ μὲν οὔτε ἀντανήγοντο οὔτε ἃ διενοήθησαν, φάρξαι τοὺς ἔσπλους, ἔτυχον ποιήσαντες, ἡσυχάζοντες δ' ἐν τῇ γῇ τάς τε ναῦς ἐπλήρουν καὶ παρεσκευάζοντο, ἢν ἐσπλέῃ τις, ὡς ἐν τῷ λιμένι ὄντι οὐ σμικρῷ ναυμαχήσοντες.

XIV. Victory of the Athenians over the Spartan fleet.
The Spartan force on Sphacteria is cut off.

Οἱ δ' Ἀθηναῖοι γνόντες καθ' ἑκάτερον τὸν ἔσπλουν ὥρμησαν ἐπ' αὐτούς, καὶ τὰς μὲν πλείους καὶ μετεώρους ἤδη τῶν νεῶν καὶ ἀντιπρῴρους προσπεσόντες ἐς φυγὴν κατέστησαν, καὶ ἐπιδιώκοντες ὡς διὰ βραχέος ἔτρωσαν μὲν πολλάς, πέντε δ' ἔλαβον καὶ μίαν τούτων αὐτοῖς ἀνδράσι· ταῖς δὲ λοιπαῖς ἐν τῇ γῇ καταπεφευγυίαις ἐνέβαλλον· αἱ δὲ καὶ πληρούμεναι ἔτι πρὶν ἀνάγεσθαι ἐκόπτοντο· καί τινας καὶ ἀναδούμενοι κενὰς εἷλκον τῶν ἀνδρῶν ἐς φυγὴν ὡρμημένων. ἃ ὁρῶντες οἱ Λακεδαιμόνιοι καὶ 2 περιαλγοῦντες τῷ πάθει, † ὅτιπερ † αὐτῶν οἱ ἄνδρες ἀπελαμβάνοντο ἐν τῇ νήσῳ, παρεβοήθουν, καὶ ἐπεσβαίνοντες ἐς τὴν θάλασσαν ξὺν τοῖς ὅπλοις ἀνθεῖλκον ἐπιλαμβανόμενοι τῶν νεῶν, καὶ ἐν τούτῳ κεκωλῦσθαι ἐδόκει ἕκαστος ᾧ μή τινι καὶ αὐτὸς ἔργῳ παρῆν. ἐγένετό τε ὁ θόρυβος μέγας καὶ ἀντηλλαγ- 3 μένος τοῦ ἑκατέρων τρόπου περὶ τὰς ναῦς· οἵ τε γὰρ Λακεδαιμόνιοι ὑπὸ προθυμίας καὶ ἐκπλήξεως ὡς εἰπεῖν ἄλλο οὐδὲν ἢ ἐκ γῆς ἐναυμάχουν, οἵ τε Ἀθηναῖοι κρατοῦντες καὶ βουλόμενοι τῇ παρούσῃ τύχῃ ὡς ἐπὶ πλεῖστον ἐπεξελθεῖν ἀπὸ νεῶν ἐπεζομάχουν. πολύν 4 τε πόνον παρασχόντες ἀλλήλοις καὶ τραυματίσαντες διεκρίθησαν, καὶ οἱ Λακεδαιμόνιοι τὰς κενὰς ναῦς πλὴν τῶν τὸ πρῶτον ληφθεισῶν διέσωσαν. κατα- 5 στάντες δὲ ἑκάτεροι ἐς τὸ στρατόπεδον οἱ μὲν τροπαῖόν τε ἔστησαν καὶ νεκροὺς ἀπέδοσαν καὶ ναυαγίων ἐκράτησαν, καὶ τὴν νῆσον εὐθὺς περιέπλεον καὶ ἐν φυλακῇ εἶχον, ὡς τῶν ἀνδρῶν ἀπειλημμένων· οἱ δ' ἐν τῇ ἠπείρῳ Πελοποννήσιοι καὶ ἀπὸ πάντων ἤδη βεβοηθηκότες ἔμενον κατὰ χώραν ἐπὶ τῇ Πύλῳ.

XV. Alarm at Sparta. An armistice at Pylus.

Ἐς δὲ τὴν Σπάρτην ὡς ἠγγέλθη τὰ γεγενημένα περὶ Πύλον, ἔδοξεν αὐτοῖς ὡς ἐπὶ ξυμφορᾷ μεγάλῃ τὰ τέλη καταβάντας ἐς τὸ στρατόπεδον βουλεύειν
2 παραχρῆμα ὁρῶντας ὅ τι ἂν δοκῇ. καὶ ὡς εἶδον ἀδύνατον ὂν τιμωρεῖν τοῖς ἄνδρασι καὶ κινδυνεύειν οὐκ ἐβούλοντο ἢ ὑπὸ λιμοῦ τι παθεῖν αὐτοὺς ἢ ὑπὸ πλήθους βιασθέντας κρατηθῆναι, ἔδοξεν αὐτοῖς πρὸς τοὺς στρατηγοὺς τῶν Ἀθηναίων, ἢν ἐθέλωσι, σπονδὰς ποιησαμένους τὰ περὶ Πύλον, ἀποστεῖλαι ἐς τὰς Ἀθήνας πρέσβεις περὶ ξυμβάσεως, καὶ τοὺς ἄνδρας ὡς τάχιστα πειρᾶσθαι κομίσασθαι.

XVI. Terms of the armistice. A Spartan embassy is sent to Athens.

Δεξαμένων δὲ τῶν στρατηγῶν τὸν λόγον ἐγίγνοντο σπονδαὶ τοιαίδε, Λακεδαιμονίους μὲν τὰς ναῦς ἐν αἷς ἐναυμάχησαν καὶ τὰς ἐν τῇ Λακωνικῇ πάσας, ὅσαι ἦσαν μακραί, παραδοῦναι κομίσαντας ἐς Πύλον Ἀθηναίοις, καὶ ὅπλα μὴ ἐπιφέρειν τῷ τειχίσματι μήτε κατὰ γῆν μήτε κατὰ θάλασσαν, Ἀθηναίους δὲ τοῖς ἐν τῇ νήσῳ ἀνδράσι σῖτον ἐᾶν τοὺς ἐν τῇ ἠπείρῳ Λακεδαιμονίους ἐσπέμπειν τακτὸν καὶ μεμαγμένον, δύο χοίνικας ἑκάστῳ Ἀττικὰς ἀλφίτων καὶ δύο κοτύλας οἴνου καὶ κρέας, θεράποντι δὲ τούτων ἡμίσεα· ταῦτα δὲ ὁρώντων τῶν Ἀθηναίων ἐσπέμπειν καὶ πλοῖον μηδὲν ἐσπλεῖν λάθρα· φυλάσσειν δὲ καὶ τὴν νῆσον Ἀθηναίους μηδὲν ἧσσον, ὅσα μὴ ἀποβαίνοντας, καὶ ὅπλα μὴ ἐπιφέρειν τῷ Πελοποννησίων στρατῷ
2 μήτε κατὰ γῆν μήτε κατὰ θάλασσαν. ὅ τι δ' ἂν τούτων παραβαίνωσιν ἑκάτεροι καὶ ὁτιοῦν, τότε

λελύσθαι τὰς σπονδάς. ἐσπεῖσθαι δὲ αὐτὰς μέχρι οὗ ἐπανέλθωσιν οἱ ἐκ τῶν Ἀθηνῶν Λακεδαιμονίων πρέσβεις· ἀποστεῖλαι δὲ αὐτοὺς τριήρει Ἀθηναίους καὶ πάλιν κομίσαι. ἐλθόντων δὲ τάς τε σπονδὰς λελύσθαι ταύτας καὶ τὰς ναῦς ἀποδοῦναι Ἀθηναίους ὁμοίας οἵασπερ ἂν παραλάβωσιν. αἱ μὲν σπονδαὶ 3 ἐπὶ τούτοις ἐγένοντο, καὶ αἱ νῆες παρεδόθησαν οὖσαι περὶ ἑξήκοντα, καὶ οἱ πρέσβεις ἀπεστάλησαν. ἀφικόμενοι δὲ ἐς τὰς Ἀθήνας ἔλεξαν τοιάδε.

XVII. Speech of the Spartan ambassadors.

"ἜΠΕΜΨΑΝ ἡμᾶς Λακεδαιμόνιοι, ὦ Ἀθηναῖοι, περὶ τῶν ἐν τῇ νήσῳ ἀνδρῶν πράξοντας ὅ τι ἂν ὑμῖν τε ὠφέλιμον ὂν τὸ αὐτὸ πείθωμεν καὶ ἡμῖν ἐς τὴν ξυμφορὰν ὡς ἐκ τῶν παρόντων κόσμον μάλιστα μέλλῃ οἴσειν. τοὺς δὲ λόγους μακροτέρους οὐ παρὰ 2 τὸ εἰωθὸς μηκυνοῦμεν, ἀλλ' ἐπιχώριον ὂν ἡμῖν οὗ μὲν βραχεῖς ἀρκῶσι μὴ πολλοῖς χρῆσθαι, πλείοσι δὲ ἐν ᾧ ἂν καιρὸς ᾖ διδάσκοντάς τι τῶν προὔργου λόγοις τὸ δέον πράσσειν. λάβετε δὲ αὐτοὺς μὴ πολεμίως 3 μηδ' ὡς ἀξύνετοι διδασκόμενοι, ὑπόμνησιν δὲ τοῦ καλῶς βουλεύσασθαι πρὸς εἰδότας ἡγησάμενοι. ὑμῖν 4 γὰρ εὐτυχίαν τὴν παροῦσαν ἔξεστι καλῶς θέσθαι, ἔχουσι μὲν ὧν κρατεῖτε, προσλαβοῦσι δὲ τιμὴν καὶ δόξαν, καὶ μὴ παθεῖν ὅπερ οἱ ἀήθως τι ἀγαθὸν λαμβάνοντες τῶν ἀνθρώπων· ἀεὶ γὰρ τοῦ πλέονος ἐλπίδι ὀρέγονται διὰ τὸ καὶ τὰ παρόντα ἀδοκήτως εὐτυχῆσαι. οἷς δὲ πλεῖσται μεταβολαὶ ἐπ' ἀμφότερα ξυμβεβήκασι, 5 δίκαιοί εἰσι καὶ ἀπιστότατοι εἶναι ταῖς εὐπραγίαις, ὃ τῇ τε ὑμετέρᾳ πόλει δι' ἐμπειρίαν καὶ ἡμῖν μάλιστ' ἂν ἐκ τοῦ εἰκότος προσείη.

XVIII. Speech continued.

"Γνῶτε δὲ καὶ ἐς τὰς ἡμετέρας νῦν ξυμφορὰς ἀπιδόντες, οἵτινες ἀξίωμα μέγιστον τῶν Ἑλλήνων ἔχοντες ἥκομεν παρ' ὑμᾶς, πρότερον αὐτοὶ κυριώτεροι νομίζοντες εἶναι δοῦναι ἐφ' ἃ νῦν ἀφιγμένοι ὑμᾶς
2 αἰτούμεθα. καίτοι οὔτε δυνάμεως ἐνδείᾳ ἐπάθομεν αὐτὸ οὔτε μείζονος προσγενομένης ὑβρίσαντες, ἀπὸ δὲ τῶν ἀεὶ ὑπαρχόντων γνώμῃ σφαλέντες, ἐν ᾧ πᾶσι
3 τὸ αὐτὸ ὁμοίως ὑπάρχει. ὥστε οὐκ εἰκὸς ὑμᾶς διὰ τὴν παροῦσαν νῦν ῥώμην πόλεώς τε καὶ τῶν προσγεγενημένων καὶ τὸ τῆς τύχης οἴεσθαι ἀεὶ μεθ' ὑμῶν
4 ἔσεσθαι. σωφρόνων δὲ ἀνδρῶν οἵτινες τἀγαθὰ ἐς ἀμφίβολον ἀσφαλῶς ἔθεντο (καὶ ταῖς ξυμφοραῖς οἱ αὐτοὶ εὐξυνετώτερον ἂν προσφέροιντο), τόν τε πόλεμον νομίσωσι μὴ καθ' ὅσον ἄν τις αὐτοῦ μέρος βούληται μεταχειρίζειν, τούτῳ ξυνεῖναι, ἀλλ' ὡς ἂν αἱ τύχαι αὐτῶν ἡγήσωνται· καὶ ἐλάχιστ' ἂν οἱ τοιοῦτοι πταίοντες διὰ τὸ μὴ τῷ ὀρθουμένῳ αὐτοῦ πιστεύοντες ἐπαίρεσθαι ἐν τῷ εὐτυχεῖν ἂν μάλιστα
5 καταλύοιντο. ὃ νῦν ὑμῖν, ὦ Ἀθηναῖοι, καλῶς ἔχει πρὸς ἡμᾶς πρᾶξαι, καὶ μήποτε ὕστερον, ἢν ἄρα μὴ πειθόμενοι σφαλῆτε, ἃ πολλὰ ἐνδέχεται, νομισθῆναι τύχῃ καὶ τὰ νῦν προχωρήσαντα κρατῆσαι, ἐξὸν ἀκίνδυνον δόκησιν ἰσχύος καὶ ξυνέσεως ἐς τὸ ἔπειτα καταλιπεῖν.

XIX. Speech continued.

"Λακεδαιμόνιοι δὲ ὑμᾶς προκαλοῦνται ἐς σπονδὰς καὶ διάλυσιν πολέμου, διδόντες μὲν εἰρήνην καὶ ξυμμαχίαν, καὶ ἄλλην φιλίαν πολλὴν καὶ οἰκειότητα ἐς ἀλλήλους ὑπάρχειν, ἀνταιτοῦντες δὲ τοὺς ἐκ τῆς νήσου ἄνδρας, καὶ ἄμεινον ἡγούμενοι ἀμφοτέροις μὴ διακινδυνεύεσθαι, εἴτε βίᾳ διαφύγοιεν παρατυχούσης τινὸς σωτηρίας εἴτε καὶ ἐκπολιορκηθέντες μᾶλλον ἂν χειρωθεῖεν. νομίζομέν τε τὰς μεγάλας ἔχθρας μάλιστ' 2 ἂν διαλύεσθαι βεβαίως, οὐκ ἢν ἀνταμυνόμενός τις καὶ ἐπικρατήσας τὰ πλείω τοῦ πολεμίου κατ' ἀνάγκην ὅρκοις ἐγκαταλαμβάνων μὴ ἀπὸ τοῦ ἴσου ξυμβῇ, ἀλλ' ἢν παρὸν τὸ αὐτὸ δρᾶσαι πρὸς τὸ ἐπιεικὲς καὶ ἀρετῇ αὐτὸν νικήσας, παρὰ ἃ προσεδέχετο, μετρίως ξυναλλαγῇ. ὀφείλων γὰρ ἤδη ὁ ἐναντίος μὴ ἀνταμύ- 3 νεσθαι ὡς βιασθεὶς ἀλλ' ἀνταποδοῦναι ἀρετήν, ἑτοιμότερός ἐστιν αἰσχύνῃ ἐμμένειν οἷς ξυνέθετο. καὶ μᾶλλον πρὸς τοὺς μειζόνως ἐχθροὺς τοῦτο δρῶσιν 4 οἱ ἄνθρωποι ἢ πρὸς τοὺς τὰ μέτρια διενεχθέντας· πεφύκασί τε τοῖς μὲν ἑκουσίως ἐνδοῦσιν ἀνθησσᾶσθαι μεθ' ἡδονῆς, πρὸς δὲ τὰ ὑπεραυχοῦντα καὶ παρὰ γνώμην διακινδυνεύειν.

XX. Speech continued.

"Ἡμῖν δὲ καλῶς, εἴπερ ποτέ, ἔχει ἀμφοτέροις ἡ ξυναλλαγή, πρίν τι ἀνήκεστον διὰ μέσου γενόμενον ἡμᾶς καταλαβεῖν, ἐν ᾧ ἀνάγκη ἀΐδιον ὑμῖν ἔχθραν πρὸς τῇ κοινῇ καὶ ἰδίαν ἔχειν, ὑμᾶς δὲ στερηθῆναι ὧν νῦν προκαλούμεθα. ἔτι δ' ὄντων ἀκρίτων, καὶ 2 ὑμῖν μὲν δόξης καὶ ἡμετέρας φιλίας προσγιγνομένης, ἡμῖν δὲ πρὸ αἰσχροῦ τινὸς < τῆς > ξυμφορᾶς μετρίως κατατιθεμένης διαλλαγῶμεν, καὶ αὐτοί τε ἀντὶ πολέμου εἰρήνην ἑλώμεθα καὶ τοῖς ἄλλοις Ἕλλησιν ἀνάπαυσιν κακῶν ποιήσωμεν· οἳ καὶ ἐν τούτῳ ὑμᾶς αἰτιωτέρους ἡγήσονται. πολεμοῦνται μὲν γὰρ ἀσαφῶς ὁποτέρων ἀρξάντων· καταλύσεως δὲ γιγνομένης, ἧς

νῦν ὑμεῖς τὸ πλέον κύριοί ἐστε, τὴν χάριν ὑμῖν
3 προσθήσουσιν. ἤν τε γνῶτε, Λακεδαιμονίοις ἔξεστιν
ὑμῖν φίλους γενέσθαι βεβαίως, αὐτῶν τε προκαλεσα-
4 μένων χαρισαμένοις τε μᾶλλον ἢ βιασαμένοις. καὶ
ἐν τούτῳ τὰ ἐνόντα ἀγαθὰ σκοπεῖτε ὅσα εἰκὸς εἶναι·
ἡμῶν γὰρ καὶ ὑμῶν ταὐτὰ λεγόντων τό γε ἄλλο
Ἑλληνικὸν ἴστε ὅτι ὑποδεέστερον ὂν τὰ μέγιστα
τιμήσει."

XXI. Terms demanded by the Athenians, on the advice of Cleon.

Οἱ μὲν οὖν Λακεδαιμόνιοι τοσαῦτα εἶπον, νομίζοντες τοὺς Ἀθηναίους ἐν τῷ πρὶν χρόνῳ σπονδῶν
μὲν ἐπιθυμεῖν, σφῶν δὲ ἐναντιουμένων κωλύεσθαι,
διδομένης δὲ εἰρήνης ἀσμένους δέξεσθαί τε καὶ τοὺς
2 ἄνδρας ἀποδώσειν. οἱ δὲ τὰς μὲν σπονδάς, ἔχοντες
τοὺς ἄνδρας ἐν τῇ νήσῳ, ἤδη σφίσιν ἐνόμιζον ἑτοίμους
εἶναι, ὁπόταν βούλωνται ποιεῖσθαι πρὸς αὐτούς, τοῦ
3 δὲ πλέονος ὠρέγοντο. μάλιστα δὲ αὐτοὺς ἐνῆγε
Κλέων ὁ Κλεαινέτου, ἀνὴρ δημαγωγὸς κατ' ἐκεῖνον
τὸν χρόνον ὢν καὶ τῷ πλήθει πιθανώτατος· καὶ
ἔπεισεν ἀποκρίνασθαι ὡς χρὴ τὰ μὲν ὅπλα καὶ
σφᾶς αὐτοὺς τοὺς ἐν τῇ νήσῳ παραδόντας πρῶτον
κομίσασθαι Ἀθήναζε, ἐλθόντων δέ, ἀποδόντας Λακεδαιμονίους Νίσαιαν καὶ Πηγὰς καὶ Τροιζῆνα καὶ
Ἀχαΐαν, ἃ οὐ πολέμῳ ἔλαβον ἀλλ' ἀπὸ τῆς προτέρας
ξυμβάσεως, Ἀθηναίων ξυγχωρησάντων κατὰ ξυμφορὰς καὶ ἐν τῷ τότε δεομένων τι μᾶλλον σπονδῶν,
κομίσασθαι τοὺς ἄνδρας καὶ σπονδὰς ποιήσασθαι
ὁπόσον ἂν δοκῇ χρόνου ἀμφοτέροις.

XXII. Negotiations are broken off.

Οἱ δὲ πρὸς μὲν τὴν ἀπόκρισιν οὐδὲν ἀντεῖπον, ξυνέδρους δὲ σφίσιν ἐκέλευον ἑλέσθαι, οἵτινες λέγοντες καὶ ἀκούοντες περὶ ἑκάστου ξυμβήσονται κατὰ ἡσυχίαν ὅ τι ἂν πείθωσιν ἀλλήλους. Κλέων δὲ 2 ἐνταῦθα δὴ πολὺς ἐνέκειτο, λέγων γιγνώσκειν μὲν καὶ πρότερον οὐδὲν ἐν νῷ ἔχοντας δίκαιον αὐτούς, σαφὲς δ' εἶναι καὶ νῦν, οἵτινες τῷ μὲν πλήθει οὐδὲν ἐθέλουσιν εἰπεῖν, ὀλίγοις δὲ ἀνδράσι ξύνεδροι βούλονται γίγνεσθαι· ἀλλὰ εἴ τι ὑγιὲς διανοοῦνται, λέγειν ἐκέλευσεν ἅπασιν. ὁρῶντες δὲ οἱ Λακεδαιμόνιοι οὔτε σφίσιν 3 οἷόν τε ὂν ἐν πλήθει εἰπεῖν, εἴ τι καὶ ὑπὸ τῆς ξυμφορᾶς ἐδόκει αὐτοῖς ξυγχωρεῖν, μὴ ἐς τοὺς ξυμμάχους διαβληθῶσιν εἰπόντες καὶ οὐ τυχόντες, οὔτε τοὺς Ἀθηναίους ἐπὶ μετρίοις ποιήσοντας ἃ προυκαλοῦντο, ἀνεχώρησαν ἐκ τῶν Ἀθηνῶν ἄπρακτοι.

XXIII. End of the armistice. The Athenians keep the Spartan fleet.

Ἀφικομένων δὲ αὐτῶν διελέλυντο εὐθὺς αἱ σπονδαὶ αἱ περὶ Πύλον, καὶ τὰς ναῦς οἱ Λακεδαιμόνιοι ἀπῄτουν, καθάπερ ξυνέκειτο· οἱ δ' Ἀθηναῖοι ἐγκλήματα ἔχοντες ἐπιδρομήν τε τῷ τειχίσματι παράσπονδον καὶ ἄλλα οὐκ ἀξιόλογα δοκοῦντα εἶναι οὐκ ἀπεδίδοσαν, ἰσχυριζόμενοι ὅτι δὴ εἴρητο, ἐὰν καὶ ὁτιοῦν παραβαθῇ, λελύσθαι τὰς σπονδάς. οἱ δὲ Λακεδαιμόνιοι ἀντέλεγόν τε, καὶ ἀδίκημα ἐπικαλέσαντες τὸ τῶν νεῶν ἀπελθόντες ἐς πόλεμον καθίσταντο. καὶ τὰ περὶ Πύλον ὑπ' 2 ἀμφοτέρων κατὰ κράτος ἐπολεμεῖτο, Ἀθηναῖοι μὲν δυοῖν νεοῖν ἐναντίαιν ἀεὶ τὴν νῆσον περιπλέοντες τῆς ἡμέρας, (τῆς δὲ νυκτὸς καὶ ἅπασαι περιώρμουν, πλὴν τὰ πρὸς τὸ πέλαγος ὁπότε ἄνεμος εἴη· καὶ ἐκ τῶν Ἀθηνῶν αὐτοῖς εἴκοσι νῆες ἀφίκοντο ἐς τὴν φυλακήν, ὥστε αἱ πᾶσαι ἑβδομήκοντα ἐγένοντο·) Πελοποννήσιοι δὲ ἐν τῇ ἠπείρῳ στρατοπεδευόμενοι καὶ προσβολὰς ποιούμενοι τῷ τείχει, σκοποῦντες καιρὸν εἴ τις παραπέσοι ὥστε τοὺς ἄνδρας σῶσαι.

XXIV. Events in Sicily. The Syracusans and Locrians prepare to fight the Athenians at sea.

Ἐν τούτῳ δὲ οἱ [ἐν τῇ Σικελίᾳ] Συρακόσιοι καὶ οἱ ξύμμαχοι, πρὸς ταῖς ἐν Μεσσήνῃ φρουρούσαις ναυσὶ τὸ ἄλλο ναυτικὸν ὃ παρεσκευάζοντο προσκομί-
2 σαντες, τὸν πόλεμον ἐποιοῦντο ἐκ τῆς Μεσσήνης (καὶ μάλιστα ἐνῆγον οἱ Λοκροὶ τῶν Ῥηγίνων κατὰ ἔχθραν, καὶ αὐτοὶ δὲ ἐσεβεβλήκεσαν πανδημεὶ ἐς τὴν γῆν αὐτῶν) καὶ ναυμαχίας ἀποπειρᾶσθαι ἐβού-
3 λοντο, ὁρῶντες τοῖς Ἀθηναίοις τὰς μὲν παρούσας ὀλίγας ναῦς, ταῖς δὲ πλείοσι καὶ μελλούσαις ἥξειν
4 πυνθανόμενοι τὴν νῆσον πολιορκεῖσθαι. εἰ γὰρ κρατήσειαν τῷ ναυτικῷ, τὸ Ῥήγιον ἤλπιζον πεζῇ τε καὶ ναυσὶν ἐφορμοῦντες ῥᾳδίως χειρώσεσθαι, καὶ ἤδη σφῶν ἰσχυρὰ τὰ πράγματα γίγνεσθαι· ξύνεγγυς γὰρ κειμένου τοῦ τε Ῥηγίου ἀκρωτηρίου τῆς Ἰταλίας τῆς τε Μεσσήνης τῆς Σικελίας, τοῖς Ἀθηναίοις [τε] οὐκ
5 ἂν εἶναι ἐφορμεῖν καὶ τοῦ πορθμοῦ κρατεῖν. ἔστι δὲ ὁ πορθμὸς ἡ μεταξὺ Ῥηγίου θάλασσα καὶ Μεσσήνης, ᾗπερ βραχύτατον Σικελία τῆς ἠπείρου ἀπέχει· καὶ ἔστιν ἡ Χάρυβδις κληθεῖσα τοῦτο, ᾗ Ὀδυσσεὺς λέγεται διαπλεῦσαι· διὰ στενότητα δὲ καὶ ἐκ μεγάλων πελαγῶν, τοῦ τε Τυρσηνικοῦ καὶ τοῦ Σικελικοῦ, ἐσπίπτουσα ἡ θάλασσα ἐς αὐτὸ καὶ ῥοώδης οὖσα εἰκότως χαλεπὴ ἐνομίσθη.

XXV. Naval engagements off the coast of Sicily. The Messanians attack Naxus, but are routed. Failure of attack made by the Leontines on Messana.

Ἐν τούτῳ οὖν τῷ μεταξὺ οἱ Συρακόσιοι καὶ οἱ ξύμμαχοι ναυσὶν ὀλίγῳ πλείοσιν ἢ τριάκοντα ἠναγκάσθησαν ὀψὲ τῆς ἡμέρας ναυμαχῆσαι περὶ πλοίου διαπλέοντος, ἀντεπαναγόμενοι πρός τε Ἀθηναίων ναῦς ἑκκαίδεκα καὶ Ῥηγίνας ὀκτώ. καὶ 2 νικηθέντες ὑπὸ τῶν Ἀθηναίων διὰ τάχους ἀπέπλευσαν ὡς ἕκαστοι ἔτυχον ἐς τὰ οἰκεῖα στρατόπεδα, τό τε ἐν τῇ Μεσσήνῃ καὶ ἐν τῷ Ῥηγίῳ, μίαν ναῦν ἀπολέσαντες· καὶ νὺξ ἐπεγένετο τῷ ἔργῳ. μετὰ δὲ τοῦτο οἱ 3 μὲν Λοκροὶ ἀπῆλθον ἐκ τῆς Ῥηγίνων, ἐπὶ δὲ τὴν Πελωρίδα τῆς Μεσσήνης ξυλλεγεῖσαι αἱ τῶν Συρακοσίων καὶ ξυμμάχων νῆες ὥρμουν καὶ ὁ πεζὸς αὐτοῖς παρῆν. προσπλεύσαντες δὲ οἱ Ἀθηναῖοι καὶ 4 Ῥηγῖνοι ὁρῶντες τὰς ναῦς κενὰς ἐνέβαλον, καὶ χειρὶ σιδηρᾷ ἐπιβληθείσῃ μίαν ναῦν αὐτοὶ ἀπώλεσαν τῶν ἀνδρῶν ἀποκολυμβησάντων. καὶ μετὰ τοῦτο τῶν 5 Συρακοσιων ἐσβάντων ἐς τὰς ναῦς καὶ παραπλεόντων ἀπὸ κάλω ἐς τὴν Μεσσήνην, αὖθις προσβαλόντες οἱ Ἀθηναῖοι, ἀποσιμωσάντων ἐκείνων καὶ προεμβαλόντων, ἑτέραν ναῦν ἀπολλύουσι. καὶ ἐν τῷ παράπλῳ 6 καὶ τῇ ναυμαχίᾳ τοιουτοτρόπῳ γενομένῃ οὐκ ἔλασσον ἔχοντες οἱ Συρακοσιοι παρεκομίσθησαν ἐς τὸν ἐν τῇ Μεσσήνῃ λιμένα.

Καὶ οἱ μὲν Ἀθηναῖοι, Καμαρίνης ἀγγελθείσης 7 προδίδοσθαι Συρακοσιοις ὑπ' Ἀρχίου καὶ τῶν μετ' αὐτοῦ, ἔπλευσαν ἐκεῖσε· Μεσσηνιοι δ' ἐν τούτῳ πανδημεὶ κατὰ γῆν καὶ ταῖς ναυσὶν ἅμα ἐστράτευσαν

8 ἐπ' Νάξον τὴν Χαλκιδικὴν ὅμορον οὖσαν. καὶ τῇ πρώτῃ ἡμέρᾳ τειχήρεις ποιήσαντες τοὺς Ναξίους ἐδῄουν τὴν γῆν, τῇ δ' ὑστεραίᾳ ταῖς μὲν ναυσὶ περιπλεύσαντες κατὰ τὸν Ἀκεσίνην ποταμὸν τὴν γῆν ἐδῄουν, τῷ δὲ πεζῷ πρὸς τὴν πόλιν προσέβαλλον. 9 ἐν τούτῳ δὲ οἱ Σικελοὶ ὑπὲρ τῶν ἄκρων πολλοὶ κατέβαινον βοηθοῦντες ἐπὶ τοὺς Μεσσηνίους. καὶ οἱ Νάξιοι ὡς εἶδον, θαρσήσαντες καὶ παρακελευόμενοι ἐν ἑαυτοῖς ὡς οἱ Λεοντῖνοι σφίσι καὶ οἱ ἄλλοι Ἕλληνες ξύμμαχοι ἐς τιμωρίαν ἐπέρχονται, ἐκδραμόντες ἄφνω ἐκ τῆς πόλεως προσπίπτουσι τοῖς Μεσσηνίοις, καὶ τρέψαντες ἀπέκτεινάν τε ὑπὲρ χιλίους καὶ οἱ λοιποὶ χαλεπῶς ἀπεχώρησαν ἐπ' οἴκου· καὶ γὰρ οἱ βάρβαροι ἐν ταῖς ὁδοῖς ἐπιπεσόντες 10 τοὺς πλείστους διέφθειραν. καὶ αἱ νῆες σχοῦσαι ἐς τὴν Μεσσήνην ὕστερον ἐπ' οἴκου ἕκασται διεκρίθησαν. Λεοντῖνοι δὲ εὐθὺς καὶ οἱ ξύμμαχοι μετὰ Ἀθηναίων ἐς τὴν Μεσσήνην ὡς κεκακωμένην ἐστράτευον, καὶ προσβάλλοντες οἱ μὲν Ἀθηναῖοι κατὰ τὸν λιμένα ταῖς ναυσὶν ἐπείρων, ὁ δὲ πεζὸς πρὸς τὴν πόλιν. 11 ἐπεκδρομὴν δὲ ποιησάμενοι οἱ Μεσσήνιοι καὶ Λοκρῶν τινὲς μετὰ τοῦ Δημοτέλους, οἳ μετὰ τὸ πάθος ἐγκατελείφθησαν φρουροί, ἐξαπιναίως προσπεσόντες τρέπουσι τοῦ στρατεύματος τῶν Λεοντίνων τὸ πολὺ καὶ ἀπέκτειναν πολλούς. ἰδόντες δὲ οἱ Ἀθηναῖοι καὶ ἀποβάντες ἀπὸ τῶν νεῶν ἐβοήθουν, καὶ κατεδίωξαν τοὺς Μεσσηνίους πάλιν ἐς τὴν πόλιν, τεταραγμένοις ἐπιγενόμενοι· καὶ τροπαῖον στήσαντες 12 ἀνεχώρησαν ἐς τὸ Ῥήγιον. μετὰ δὲ τοῦτο οἱ μὲν ἐν τῇ Σικελίᾳ Ἕλληνες ἄνευ τῶν Ἀθηναίων κατὰ γῆν ἐστράτευον ἐπ' ἀλλήλους.

XXVI. Athenian difficulties at Pylus. The Spartans on Sphacteria hold out.

Ἐν δὲ τῇ Πύλῳ ἔτι ἐπολιόρκουν τοὺς ἐν τῇ νήσῳ Λακεδαιμονίους οἱ Ἀθηναῖοι, καὶ τὸ ἐν τῇ ἠπείρῳ στρατόπεδον τῶν Πελοποννησίων κατὰ χώραν ἔμενεν. ἐπίπονος δ' ἦν τοῖς Ἀθηναίοις ἡ φυλακὴ σίτου τε 2 ἀπορίᾳ καὶ ὕδατος· οὐ γὰρ ἦν κρήνη ὅτι μὴ μία ἐν αὐτῇ τῇ ἀκροπόλει τῆς Πύλου καὶ αὕτη οὐ μεγάλη, ἀλλὰ διαμώμενοι τὸν κάχληκα οἱ πλεῖστοι ἐπὶ τῇ θαλάσσῃ ἔπινον οἷον εἰκὸς ὕδωρ. στενοχωρία τε ἐν 3 ὀλίγῳ στρατοπεδευομένοις ἐγίγνετο, καὶ τῶν νεῶν οὐκ ἐχουσῶν ὅρμον αἱ μὲν σῖτον ἐν τῇ γῇ ᾑροῦντο κατὰ μέρος, αἱ δὲ μετέωροι ὥρμουν. ἀθυμίαν τε 4 πλείστην ὁ χρόνος παρεῖχε παρὰ λόγον ἐπιγιγνόμενος, οὓς ᾤοντο ἡμερῶν ὀλίγων ἐκπολιορκήσειν ἐν νήσῳ τε ἐρήμῃ καὶ ὕδατι ἁλμυρῷ χρωμένους. αἴτιον δὲ ἦν οἱ 5 Λακεδαιμόνιοι προειπόντες ἐς τὴν νῆσον ἐσάγειν σῖτόν τε τὸν βουλόμενον ἀληλεμένον καὶ οἶνον καὶ τυρὸν καὶ εἴ τι ἄλλο βρῶμα, οἷα ἂν ἐς πολιορκίαν ξυμφέρῃ, τάξαντες ἀργυρίου πολλοῦ καὶ τῶν Εἱλώτων τῷ ἐσαγαγόντι ἐλευθερίαν ὑπισχνούμενοι. καὶ ἐσῆγον 6 ἄλλοι τε παρακινδυνεύοντες καὶ μάλιστα οἱ Εἵλωτες, ἀπαίροντες ἀπὸ τῆς Πελοποννήσου ὁπόθεν τύχοιεν καὶ καταπλέοντες ἔτι νυκτὸς ἐς τὰ πρὸς τὸ πέλαγος τῆς νήσου. μάλιστα δὲ ἐτήρουν ἀνέμῳ καταφέρεσθαι· 7 ῥᾷον γὰρ τὴν φυλακὴν τῶν τριήρων ἐλάνθανον, ὁπότε πνεῦμα ἐκ πόντου εἴη· ἄπορον γὰρ ἐγίγνετο περιορμεῖν, τοῖς δὲ ἀφειδὴς ὁ κατάπλους καθειστήκει· ἐπώκελλον γὰρ τὰ πλοῖα τετιμημένα χρημάτων, καὶ οἱ ὁπλῖται περὶ τὰς κατάρσεις τῆς νήσου ἐφύλασσον. ὅσοι δὲ γαλήνῃ κινδυνεύσειαν, ἡλίσκοντο. ἐσένεον 8

δὲ καὶ κατὰ τὸν λιμένα κολυμβηταὶ ὕφυδροι, καλῳδίῳ ἐν ἀσκοῖς ἐφέλκοντες μήκωνα μεμελιτωμένην καὶ λίνου σπέρμα κεκομμένον· ὧν τὸ πρῶτον λανθανόντων
9 φυλακαὶ ὕστερον ἐγένοντο. παντί τε τρόπῳ ἑκάτεροι ἐτεχνῶντο, οἱ μὲν ἐσπέμπειν τὰ σιτία, οἱ δὲ μὴ λανθάνειν σφᾶς.

XXVII. Disappointment at Athens. Cleon criticises the Athenian commanders.

Ἐν δὲ ταῖς Ἀθήναις πυνθανόμενοι περὶ τῆς στρατιᾶς ὅτι ταλαιπωρεῖται καὶ σῖτος τοῖς ἐν τῇ νήσῳ ὅτι ἐσπλεῖ, ἠπόρουν καὶ ἐδεδοίκεσαν μὴ σφῶν χειμὼν τὴν φυλακὴν ἐπιλάβοι, ὁρῶντες τῶν τε ἐπιτηδείων τὴν περὶ τὴν Πελοπόννησον κομιδὴν ἀδύνατον ἐσομένην, ἅμα ἐν χωρίῳ ἐρήμῳ καὶ οὐδ' ἐν θέρει οἷοί τε ὄντες ἱκανὰ περιπέμπειν, τόν τε ἔφορμον χωρίων ἀλιμένων ὄντων οὐκ ἐσόμενον, ἀλλ' ἢ σφῶν ἀνέντων τὴν φυλακὴν περιγενήσεσθαι τοὺς ἄνδρας ἢ τοῖς πλοίοις ἃ τὸν σῖτον αὐτοῖς ἦγε χειμῶνα τηρήσαντας
2 ἐκπλεύσεσθαι. πάντων δὲ ἐφοβοῦντο μάλιστα τοὺς Λακεδαιμονίους, ὅτι ἔχοντάς τι ἰσχυρὸν αὐτοὺς ἐνόμιζον οὐκέτι σφίσιν ἐπικηρυκεύεσθαι· καὶ μετε-
3 μέλοντο τὰς σπονδὰς οὐ δεξάμενοι. Κλέων δὲ γνοὺς αὐτῶν τὴν ἐς αὐτὸν ὑποψίαν περὶ τῆς κωλύμης τῆς ξυμβάσεως οὐ τἀληθῆ ἔφη λέγειν τοὺς ἐξαγγέλλοντας. παραινούντων δὲ τῶν ἀφιγμένων, εἰ μὴ σφίσι πιστεύουσι, κατασκόπους τινὰς πέμψαι, ᾑρέθη κατάσκοπος
4 αὐτὸς μετὰ Θεογένους ὑπὸ Ἀθηναίων. καὶ γνοὺς ὅτι ἀναγκασθήσεται ἢ ταὐτὰ λέγειν οἷς διέβαλλεν ἢ τἀναντία εἰπὼν ψευδὴς φανήσεσθαι, παρῄνει τοῖς

Ἀθηναίοις, ὁρῶν αὐτοὺς καὶ ὡρμημένους τι τὸ πλέον τῇ γνώμῃ στρατεύειν, ὡς χρὴ κατασκόπους μὲν μὴ πέμπειν μηδὲ διαμέλλειν καιρὸν παριέντας, εἰ δὲ δοκεῖ αὐτοῖς ἀληθῆ εἶναι τὰ ἀγγελλόμενα, πλεῖν ἐπὶ τοὺς ἄνδρας. καὶ ἐς Νικίαν τὸν Νικηράτου στρατηγὸν 5 ὄντα ἀπεσήμαινεν, ἐχθρὸς ὢν καὶ ἐπιτιμῶν, ῥάδιον εἶναι παρασκευῇ, εἰ ἄνδρες εἶεν οἱ στρατηγοί, πλεύσαντας λαβεῖν τοὺς ἐν τῇ νήσῳ, καὶ αὐτός γ' ἄν, εἰ ἦρχε, ποιῆσαι τοῦτο.

XXVIII. Nicias offers the command to Cleon, who is forced to accept and promises to take Sphacteria within twenty days.

Ὁ δὲ Νικίας τῶν τε Ἀθηναίων τι ὑποθορυβησάντων ἐς τὸν Κλέωνα, ὅτι οὐ καὶ νῦν πλεῖ, εἰ ῥᾴδιόν γε αὐτῷ φαίνεται, καὶ ἅμα ὁρῶν αὐτὸν ἐπιτιμῶντα, ἐκέλευεν ἥν τινα βούλεται δύναμιν λαβόντα τὸ ἐπὶ σφᾶς εἶναι ἐπιχειρεῖν. ὁ δὲ τὸ μὲν πρῶτον οἰόμενος 2 αὐτὸν λόγῳ μόνον ἀφιέναι ἑτοῖμος ἦν, γνοὺς δὲ τῷ ὄντι παραδωσείοντα ἀνεχώρει καὶ οὐκ ἔφη αὐτὸς ἀλλ' ἐκεῖνον στρατηγεῖν, δεδιὼς ἤδη καὶ οὐκ ἂν οἰόμενός οἱ αὐτὸν τολμῆσαι ὑποχωρῆσαι. αὖθις δὲ 3 ὁ Νικίας ἐκέλευε, καὶ ἐξίστατο τῆς ἐπὶ Πύλῳ ἀρχῆς, καὶ μάρτυρας τοὺς Ἀθηναίους ἐποιεῖτο. οἱ δέ, οἷον ὄχλος φιλεῖ ποιεῖν, ὅσῳ μᾶλλον ὁ Κλέων ὑπέφευγε τὸν πλοῦν καὶ ἐξανεχώρει τὰ εἰρημένα, τόσῳ ἐπεκελεύοντο τῷ Νικίᾳ παραδιδόναι τὴν ἀρχὴν καὶ ἐκείνῳ ἐπεβόων πλεῖν. ὥστε οὐκ ἔχων ὅπως τῶν εἰρημένων 4 ἔτι ἐξαπαλλαγῇ, ὑφίσταται τὸν πλοῦν, καὶ παρελθὼν οὔτε φοβεῖσθαι ἔφη Λακεδαιμονίους πλεύσεσθαί τε λαβὼν ἐκ μὲν τῆς πόλεως οὐδένα, Λημνίους δὲ καὶ

Ἰμβρίους τοὺς παρόντας καὶ πελταστὰς οἳ ἦσαν ἔκ τε Αἴνου βεβοηθηκότες καὶ ἄλλοθεν τοξότας τετρακοσίους· ταῦτα δὲ ἔχων ἔφη πρὸς τοῖς ἐν Πύλῳ στρατιώταις ἐντὸς ἡμερῶν εἴκοσιν ἢ ἄξειν Λακεδαι-
5 μονίους ζῶντας ἢ αὐτοῦ ἀποκτενεῖν. τοῖς δὲ Ἀθηναίοις ἐνέπεσε μέν τι καὶ γέλωτος τῇ κουφολογίᾳ αὐτοῦ, ἀσμένοις δ᾽ ὅμως ἐγίγνετο τοῖς σώφροσι τῶν ἀνθρώπων, λογιζομένοις δυοῖν ἀγαθοῖν τοῦ ἑτέρου τεύξεσθαι, ἢ Κλέωνος ἀπαλλαγήσεσθαι, ὃ μᾶλλον ἤλπιζον, ἢ σφαλεῖσι γνώμης Λακεδαιμονίους σφίσι χειρώσεσθαι.

XXIX. Difficulties of an attack on Sphacteria.

Καὶ πάντα διαπραξάμενος ἐν τῇ ἐκκλησίᾳ, καὶ ψηφισαμένων Ἀθηναίων αὐτῷ τὸν πλοῦν, τῶν τε ἐν Πύλῳ στρατηγῶν ἕνα προσελόμενος Δημοσθένη,
2 τὴν ἀναγωγὴν διὰ τάχους ἐποιεῖτο. τὸν δὲ Δημοσθένη προσέλαβε πυνθανόμενος τὴν ἀπόβασιν αὐτὸν ἐς τὴν νῆσον διανοεῖσθαι. οἱ γὰρ στρατιῶται κακοπαθοῦντες τοῦ χωρίου τῇ ἀπορίᾳ καὶ μᾶλλον πολιορκούμενοι ἢ πολιορκοῦντες ὥρμηντο διακινδυνεῦσαι. καὶ αὐτῷ ἔτι
3 ῥώμην καὶ ἡ νῆσος ἐμπρησθεῖσα παρέσχε. πρότερον μὲν γὰρ οὔσης αὐτῆς ὑλώδους ἐπὶ τὸ πολὺ καὶ ἀτριβοῦς διὰ τὴν ἀεὶ ἐρημίαν ἐφοβεῖτο καὶ πρὸς τῶν πολεμίων τοῦτο ἐνόμιζε μᾶλλον εἶναι· πολλῷ γὰρ ἂν στρατοπέδῳ ἀποβάντι ἐξ ἀφανοῦς χωρίου προσβάλλοντας αὐτοὺς βλάπτειν. σφίσι μὲν γὰρ τὰς ἐκείνων ἁμαρτίας καὶ παρασκευὴν ὑπὸ τῆς ὕλης οὐκ ἂν ὁμοίως δῆλα εἶναι, τοῦ δὲ αὐτῶν στρατοπέδου καταφανῆ ἂν εἶναι πάντα τὰ ἁμαρτήματα, ὥστε προσπίπτειν ἂν αὐτοὺς ἀπροσδοκήτως ᾗ βούλοιντο· ἐπ᾽ ἐκείνοις γὰρ

ἂν εἶναι τὴν ἐπιχείρησιν. εἰ δ' αὖ ἐς δασὺ χωρίον 4 βιάζοιτο ὁμόσε ἰέναι, τοὺς ἐλάσσους ἐμπείρους δὲ τῆς χώρας κρείσσους ἐνόμιζε τῶν πλεόνων ἀπείρων· λανθάνειν τε ἂν τὸ ἑαυτῶν στρατόπεδον πολὺ ὂν διαφθειρόμενον, οὐκ οὔσης τῆς προσόψεως ᾗ χρὴ ἀλλήλοις ἐπιβοηθεῖν.

XXX. Demosthenes is encouraged to make the attack by the burning of the wood on Sphacteria. Arrival of Cleon, and offer of terms to the Spartans.

Ἀπὸ δὲ τοῦ Αἰτωλικοῦ πάθους, ὃ διὰ τὴν ὕλην μέρος τι ἐγένετο, οὐχ ἥκιστα αὐτὸν ταῦτα ἐσῄει. τῶν 2 δὲ στρατιωτῶν ἀναγκασθέντων διὰ τὴν στενοχωρίαν τῆς νήσου τοῖς ἐσχάτοις προσίσχοντας ἀριστοποιεῖσθαι διὰ προφυλακῆς καὶ ἐμπρήσαντός τινος κατὰ μικρὸν τῆς ὕλης ἄκοντος καὶ ἀπὸ τούτου πνεύματος ἐπιγενομένου τὸ πολὺ αὐτῆς ἔλαθε κατακαυθέν. οὕτω 3 δὴ τούς τε Λακεδαιμονίους μᾶλλον κατιδὼν πλείους ὄντας, ὑπονοῶν πρότερον ἐλάσσοσι τὸν σῖτον αὐτοῦ ἐσπέμπειν, τήν τε νῆσον εὐαποβατωτέραν οὖσαν, τότε ὡς ἐπ' ἀξιόχρεων τοὺς Ἀθηναίους μᾶλλον σπουδὴν ποιεῖσθαι τὴν ἐπιχείρησιν παρεσκευάζετο, στρατιάν τε μεταπέμπων ἐκ τῶν ἐγγὺς ξυμμάχων καὶ τὰ ἄλλα ἑτοιμάζων. Κλέων δὲ ἐκείνῳ τε προπέμψας 4 ἄγγελον ὡς ἥξων, καὶ ἔχων στρατιὰν ἣν ᾐτήσατο, ἀφικνεῖται ἐς Πύλον. καὶ ἅμα γενόμενοι πέμπουσι πρῶτον ἐς τὸ ἐν τῇ ἠπείρῳ στρατόπεδον κήρυκα, προκαλούμενοι, εἰ βούλοιντο, ἄνευ κινδύνου τοὺς ἐν τῇ νήσῳ ἄνδρας σφίσι τά τε ὅπλα καὶ σφᾶς αὐτοὺς κελεύειν παραδοῦναι, ἐφ' ᾧ φυλακῇ τῇ μετρίᾳ τηρήσονται, ἕως ἄν τι περὶ τοῦ πλέονος ξυμβαθῇ.

XXXI. Preparations for the attack, and disposition of the Spartan forces.

Οὐ προσδεξαμένων δὲ αὐτῶν μίαν μὲν ἡμέραν ἐπέσχον, τῇ δ' ὑστεραίᾳ ἀνηγάγοντο μὲν νυκτὸς ἐπ' ὀλίγας ναῦς τοὺς ὁπλίτας πάντας ἐπιβιβάσαντες, πρὸ δὲ τῆς ἕω ὀλίγον ἀπέβαινον τῆς νήσου ἑκατέρωθεν, ἔκ τε τοῦ πελάγους καὶ πρὸς τοῦ λιμένος, ὀκτακόσιοι μάλιστα ὄντες ὁπλῖται, καὶ ἐχώρουν δρόμῳ ἐπὶ τὸ πρῶτον φυλακτήριον τῆς νήσου. ὧδε γὰρ διετετάχατο. 2 ἐν ταύτῃ μὲν τῇ πρώτῃ φυλακῇ ὡς τριάκοντα ἦσαν ὁπλῖται, μέσον δὲ καὶ ὁμαλώτατόν τε καὶ περὶ τὸ ὕδωρ οἱ πλεῖστοι αὐτῶν και Ἐπιτάδας ὁ ἄρχων εἶχε, μέρος δέ τι οὐ πολὺ αὐτὸ τὸ ἔσχατον ἐφύλασσε τῆς νήσου τὸ πρὸς τὴν Πύλον, ὃ ἦν ἔκ τε θαλάσσης ἀπόκρημνον καὶ ἐκ τῆς γῆς ἥκιστα ἐπίμαχον· καὶ γάρ τι καὶ ἔρυμα αὐτόθι ἦν παλαιὸν λίθων λογάδην πεποιημένον, ὃ ἐνόμιζον σφίσιν ὠφέλιμον ἂν εἶναι, εἰ καταλαμβάνοι ἀναχώρησις βιαιοτέρα. οὕτω μὲν τεταγμένοι ἦσαν.

XXXII. The Athenians land, destroy the first detachment of Spartans, and prepare to attack the main body.

Οἱ δὲ Ἀθηναῖοι τοὺς μὲν πρώτους φύλακας, οἷς ἐπέδραμον, εὐθὺς διαφθείρουσιν ἔν τε ταῖς εὐναῖς ἔτι καὶ ἀναλαμβάνοντας τὰ ὅπλα, λαθόντες τὴν ἀπόβασιν, οἰομένων αὐτῶν τὰς ναῦς κατὰ τὸ ἔθος ἐς ἔφορμον τῆς 2 νυκτὸς πλεῖν. ἅμα δὲ ἕῳ γιγνομένῃ καὶ ὁ ἄλλος στρατὸς ἀπέβαινον, ἐκ μὲν νεῶν ἑβδομήκοντα καὶ ὀλίγῳ πλειόνων πάντες πλὴν θαλαμιῶν, ὡς ἕκαστοι ἐσκευασμένοι, τοξόται δὲ ὀκτακόσιοι και πελτασται

οὐκ ἐλάσσους τούτων, Μεσσηνίων τε οἱ βεβοηθηκότες
καὶ οἱ ἄλλοι ὅσοι περὶ Πύλον κατεῖχον πάντες πλὴν
τῶν ἐπὶ τοῦ τείχους φυλάκων. Δημοσθένους δὲ 3
τάξαντος διέστησαν κατὰ διακοσίους τε καὶ πλείους,
ἔστι δ' ᾗ ἐλάσσους, τῶν χωρίων τὰ μετέωρα <κατα>λα-
βόντες, ὅπως ὅτι πλείστη ἀπορία ᾖ τοῖς πολεμίοις
πανταχόθεν κεκυκλωμένοις, καὶ μὴ ἔχωσι πρὸς ὅ τι
ἀντιτάξωνται, ἀλλ' ἀμφίβολοι γίγνωνται τῷ πλήθει,
εἰ μὲν τοῖς πρόσθεν ἐπίοιεν, ὑπὸ τῶν κατόπιν βαλλό-
μενοι, εἰ δὲ τοῖς πλαγίοις, ὑπὸ τῶν ἑκατέρωθεν παρα-
τεταγμένων. κατὰ νώτου τε ἀεὶ ἔμελλον αὐτοῖς, ᾗ 4
χωρήσειαν, οἱ πολέμιοι ἔσεσθαι ψιλοὶ καὶ οἱ ἀπορώτατοι, τοξεύμασι καὶ ἀκοντίοις καὶ λίθοις καὶ σφεν-
δόναις ἐκ πολλοῦ ἔχοντες ἀλκήν, οἷς μηδὲ ἐπελθεῖν
οἷόν τε ἦν· φεύγοντές τε γὰρ ἐκράτουν καὶ ἀναχωροῦσιν ἐπέκειντο. τοιαύτη μὲν γνώμη ὁ Δημοσθένης 5
τό τε πρῶτον τὴν ἀπόβασιν ἐπενόει καὶ ἐν τῷ ἔργῳ
ἔταξεν.

XXXIII. Description of the battle. Attacks of the Athenian light-armed troops.

Οἱ δὲ περὶ τὸν Ἐπιτάδαν καὶ ὅπερ ἦν πλεῖστον
τῶν ἐν τῇ νήσῳ, ὡς εἶδον τό τε πρῶτον φυλακτήριον
διεφθαρμένον καὶ στρατὸν σφίσιν ἐπιόντα, ξυνετά-
ξαντο καὶ τοῖς ὁπλίταις τῶν Ἀθηναίων ἐπῇσαν, βουλόμενοι ἐς χεῖρας ἐλθεῖν· ἐξ ἐναντίας γὰρ οὗτοι
καθειστήκεσαν, ἐκ πλαγίου δὲ οἱ ψιλοὶ καὶ κατὰ
νώτου. τοῖς μὲν οὖν ὁπλίταις οὐκ ἐδυνήθησαν προσ- 2
μῖξαι οὐδὲ τῇ σφετέρᾳ ἐμπειρίᾳ χρήσασθαι· οἱ γὰρ
ψιλοὶ ἑκατέρωθεν βαλλοντες εἶργον, καὶ ἅμα ἐκεῖνοι

οὐκ ἀντεπῆσαν ἀλλ' ἡσύχαζον· τοὺς δὲ ψιλούς, ᾗ μάλιστα αὐτοῖς προσθέοντες προσκέοιντο, ἔτρεπον, καὶ οἳ ὑποστρέφοντες ἠμύνοντο, ἄνθρωποι κούφως τε ἐσκευασμένοι καὶ προλαμβάνοντες ῥᾳδίως τῆς φυγῆς χωρίων τε χαλεπότητι καὶ ὑπὸ τῆς πρὶν ἐρημίας τραχέων ὄντων, ἐν οἷς οἱ Λακεδαιμόνιοι οὐκ ἐδύναντο διώκειν ὅπλα ἔχοντες.

XXXIV. The Spartan force in difficulties.

Χρόνον μὲν οὖν τινὰ ὀλίγον οὕτω πρὸς ἀλλήλους ἠκροβολίσαντο· τῶν δὲ Λακεδαιμονίων οὐκέτι ὀξέως ἐπεκθεῖν ᾗ προσπίπτοιεν δυναμένων, γνόντες αὐτοὺς οἱ ψιλοὶ βραδυτέρους ἤδη ὄντας τῷ ἀμύνασθαι, καὶ αὐτοὶ τῇ τε ὄψει τοῦ θαρσεῖν τὸ πλεῖστον εἰληφότες πολλαπλάσιοι φαινόμενοι καὶ ξυνειθισμένοι μᾶλλον μηκέτι δεινοὺς αὐτοὺς ὁμοίως σφίσι φαίνεσθαι, ὅτι οὐκ εὐθὺς ἄξια τῆς προσδοκίας ἐπεπόνθεσαν, ὥσπερ ὅτε πρῶτον ἀπέβαινον τῇ γνώμῃ δεδουλωμένοι ὡς ἐπὶ Λακεδαιμονίους, καταφρονήσαντες καὶ ἐμβοήσαντες ἀθρόοι ὥρμησαν ἐπ' αὐτοὺς καὶ ἔβαλλον λίθοις τε καὶ τοξεύμασι καὶ ἀκοντίοις, ὡς ἕκαστός τι πρόχειρον
2 εἶχε. γενομένης δὲ τῆς βοῆς ἅμα τῇ ἐπιδρομῇ ἔκπληξίς τε ἐνέπεσεν ἀνθρώποις ἀήθεσι τοιαύτης μάχης καὶ ὁ κονιορτὸς τῆς ὕλης νεωστὶ κεκαυμένης ἐχώρει πολὺς ἄνω, ἄπορόν τε ἦν ἰδεῖν τὸ πρὸ αὑτοῦ ὑπὸ τῶν τοξευμάτων καὶ λίθων ἀπὸ πολλῶν ἀνθρώπων
3 μετὰ τοῦ κονιορτοῦ ἅμα φερομένων. τό τε ἔργον ἐνταῦθα χαλεπὸν τοῖς Λακεδαιμονίοις καθίστατο· οὔτε γὰρ οἱ πῖλοι ἔστεγον τὰ τοξεύματα, δοράτιά τε ἐναπεκέκλαστο βαλλομένων, εἶχόν τε οὐδὲν σφίσιν αὐτοῖς χρήσασθαι ἀποκεκλῃμένοι μὲν τῇ ὄψει τοῦ

προορᾶν, ὑπὸ δὲ τῆς μείζονος βοῆς τῶν πολεμίων τὰ ἐν αὐτοῖς παραγγελλόμενα οὐκ ἐσακούοντες, κινδύνου τε πανταχόθεν περιεστῶτος καὶ οὐκ ἔχοντες ελπίδα καθ' ὅτι χρὴ ἀμυνομένους σωθῆναι.

XXXV. Retreat of the Spartans to their last stronghold.

Τέλος δὲ τραυματιζομένων ἤδη πολλῶν διὰ τὸ ἀεὶ ἐν τῷ αὐτῷ ἀναστρέφεσθαι, ξυγκλῄσαντες ἐχώρησαν ἐς τὸ ἔσχατον ἔρυμα τῆς νήσου, ὃ οὐ πολὺ ἀπεῖχε, καὶ τοὺς ἑαυτῶν φύλακας. ὡς δὲ ἐνέδοσαν, ἐνταῦθα ἤδη 2 πολλῷ ἔτι πλέονι βοῇ τεθαρσηκότες οἱ ψιλοὶ ἐπέκειντο, καὶ τῶν Λακεδαιμονίων ὅσοι μὲν ὑποχωροῦντες ἐγκατελαμβάνοντο, ἀπέθνῃσκον, οἱ δὲ πολλοὶ διαφυγόντες ἐς τὸ ἔρυμα μετὰ τῶν ταύτῃ φυλάκων ἐτάξαντο παρὰ πᾶν ὡς ἀμυνούμενοι ᾗπερ ἦν ἐπίμαχον. καὶ οἱ 3 Ἀθηναῖοι ἐπισπόμενοι περίοδον μὲν αὐτῶν καὶ κύκλωσιν χωρίου ἰσχύι οὐκ εἶχον, προσιόντες δὲ ἐξ ἐναντίας ὤσασθαι ἐπειρῶντο. καὶ χρόνον μὲν πολὺν 4 καὶ τῆς ἡμέρας τὸ πλεῖστον ταλαιπωρούμενοι αμφότεροι ὑπό τε τῆς μάχης καὶ δίψης καὶ ἡλίου ἀντεῖχον, πειρώμενοι οἱ μὲν ἐξελάσασθαι ἐκ τοῦ μετεώρου, οἱ δὲ μὴ ἐνδοῦναι· ῥᾷον δ' οἱ Λακεδαιμόνιοι ἠμύνοντο ἢ ἐν τῷ πριν, οὐκ οὔσης σφῶν τῆς κυκλώσεως ἐς τὰ πλάγια.

XXXVI. An Athenian force goes round by the cliff and attacks the Spartans in the rear.

Ἐπειδὴ δὲ ἀπέραντον ἦν, προσελθὼν ὁ τῶν Μεσσηνίων στρατηγὸς Κλέωνι καὶ Δημοσθένει ἄλλως ἔφη πονεῖν σφᾶς· εἰ δὲ βούλονται ἑαυτῷ δοῦναι τῶν

τοξοτῶν μέρος τι καὶ τῶν ψιλῶν περιιέναι κατὰ νώτου αὐτοῖς ὁδῷ ᾗ ἂν αὐτὸς εὕρῃ, δοκεῖν βιάσεσθαι τὴν
2 ἔφοδον. λαβὼν δὲ ἃ ᾐτήσατο, ἐκ τοῦ ἀφανοῦς ὁρμήσας ὥστε μὴ ἰδεῖν ἐκείνους, κατὰ τὸ ἀεὶ παρεῖκον τοῦ κρημνώδους τῆς νήσου προβαίνων καὶ ᾗ οἱ Λακεδαιμόνιοι χωρίου ἰσχύι πιστεύσαντες οὐκ ἐφύλασσον, χαλεπῶς τε καὶ μόλις περιελθὼν ἔλαθε, καὶ ἐπὶ τοῦ μετεώρου ἐξαπίνης ἀναφανεὶς κατὰ νώτου αὐτῶν τοὺς μὲν τῷ ἀδοκήτῳ ἐξέπληξε, τοὺς δὲ ἃ προσεδέχοντο
3 ἰδόντας πολλῷ μᾶλλον ἐπέρρωσε. καὶ οἱ Λακεδαιμόνιοι βαλλόμενοί τε ἀμφοτέρωθεν ἤδη καὶ γιγνόμενοι ἐν τῷ αὐτῷ ξυμπτώματι, ὡς μικρὸν μεγάλῳ εἰκάσαι, τῷ ἐν Θερμοπύλαις, ἐκεῖνοί τε γὰρ τῇ ἀτραπῷ περιελθόντων τῶν Περσῶν διεφθάρησαν, οὗτοί τε ἀμφίβολοι ἤδη ὄντες οὐκέτι ἀντεῖχον, ἀλλὰ πολλοῖς τε ὀλίγοι μαχόμενοι καὶ ἀσθενείᾳ σωμάτων διὰ την σιτοδείαν ὑπεχώρουν, καὶ οἱ Ἀθηναῖοι ἐκράτουν ἤδη τῶν ἐφόδων.

XXXVII. Cleon and Demosthenes summon the Spartans to surrender.

Γνοὺς δὲ ὁ Κλέων καὶ ὁ Δημοσθένης, εἰ καὶ ὁποσονοῦν μᾶλλον ἐνδώσουσι, διαφθαρησομένους αὐτοὺς ὑπὸ τῆς σφετέρας στρατιᾶς, ἔπαυσαν τὴν μάχην καὶ τοὺς ἑαυτῶν ἀπεῖρξαν, βουλόμενοι ἀγαγεῖν αὐτοὺς Ἀθηναίοις ζῶντας, εἴ πως τοῦ κηρύγματος ἀκούσαντες ἐπικλασθεῖεν τῇ γνώμῃ [τὰ ὅπλα παραδοῦναι] καὶ
2 ἡσσηθεῖεν τοῦ παρόντος δεινοῦ. ἐκήρυξάν τε, εἰ βούλονται, τὰ ὅπλα παραδοῦναι καὶ σφᾶς αὐτοὺς Ἀθηναίοις ὥστε βουλεῦσαι ὅ τι ἂν ἐκείνοις δοκῇ.

XXXVIII. Surrender of the Spartans.

Οἱ δὲ ἀκούσαντες παρεῖσαν τὰς ἀσπίδας οἱ πλεῖστοι καὶ τὰς χεῖρας ἀνέσεισαν, δηλοῦντες προσιεσθαι τὰ κεκηρυγμένα. μετὰ δὲ ταῦτα γενομένης τῆς ἀνοκωχῆς ξυνῆλθον ἐς λόγους ὅ τε Κλέων καὶ ὁ Δημοσθένης καὶ ἐκείνων Στύφων ὁ Φάρακος, τῶν πρότερον ἀρχόντων τοῦ μὲν πρώτου τεθνηκότος, Ἐπιτάδου, τοῦ δὲ μετ᾽ αὐτὸν Ἱππαγρέτου ἐφῃρημένου ἐν τοῖς νεκροῖς ἔτι ζῶντος κειμένου ὡς τεθνεῶτος, αὐτὸς τρίτος ἐφῃρημένος ἄρχειν κατὰ νόμον, εἴ τι ἐκεῖνοι πάσχοιεν. ἔλεγε δὲ ὁ Στύφων καὶ οἱ μετ᾽ αὐτοῦ ὅτι βούλονται 2 διακηρυκεύσασθαι πρὸς τοὺς ἐν τῇ ἠπείρῳ Λακεδαιμονίους ὅ τι χρὴ σφᾶς ποιεῖν. καὶ ἐκείνων μὲν οὐδένα 3 ἀφιέντων, αὐτῶν δὲ τῶν Ἀθηναίων καλούντων ἐκ τῆς ἠπείρου κήρυκας καὶ γενομένων ἐπερωτήσεων δὶς ἢ τρίς, ὁ τελευταῖος διαπλεύσας αὐτοῖς ἀπὸ τῶν ἐκ τῆς ἠπείρου Λακεδαιμονίων [ἀνὴρ] ἀπήγγειλεν ὅτι "οἱ Λακεδαιμόνιοι κελεύουσιν ὑμᾶς αὐτοὺς περὶ ὑμῶν αὐτῶν βουλεύεσθαι, μηδὲν αἰσχρὸν ποιοῦντας." οἱ δὲ καθ᾽ ἑαυτοὺς βουλευσάμενοι τὰ ὅπλα παρέδοσαν καὶ σφᾶς αὐτούς. καὶ ταύτην μὲν τὴν ἡμέραν καὶ τὴν 4 ἐπιοῦσαν νύκτα ἐν φυλακῇ εἶχον αὐτοὺς οἱ Ἀθηναῖοι· τῇ δ᾽ ὑστεραίᾳ οἱ μὲν Ἀθηναῖοι τροπαῖον στήσαντες ἐν τῇ νήσῳ τὰ ἄλλα διεσκευάζοντο ὡς ἐς πλοῦν, καὶ τοὺς ἄνδρας τοῖς τριηράρχοις διεδίδοσαν ἐς φυλακήν, οἱ δὲ Λακεδαιμόνιοι κήρυκα πέμψαντες τοὺς νεκροὺς διεκομίσαντο. ἀπέθανον δ᾽ ἐν τῇ νήσῳ καὶ ζῶντες 5 ἐλήφθησαν τοσοίδε· εἴκοσι μὲν ὁπλῖται διέβησαν καὶ τετρακόσιοι οἱ πάντες· τούτων ζῶντες ἐκομίσθησαν ὀκτὼ ἀποδέοντες τριακόσιοι, οἱ δὲ ἄλλοι ἀπέθανον. καὶ Σπαρτιᾶται τούτων ἦσαν τῶν ζώντων περὶ εἴκοσι καὶ ἑκατόν. Ἀθηναίων δὲ οὐ πολλοὶ διεφθάρησαν· ἡ γὰρ μάχη οὐ σταδία ἦν.

XXXIX. Length of the siege. The Athenians and the Spartans on the mainland return homewards.

Χρόνος δὲ ὁ ξύμπας ἐγένετο, ὅσον οἱ ἄνδρες οἱ ἐν τῇ νήσῳ ἐπολιορκήθησαν ἀπὸ τῆς ναυμαχίας μέχρι τῆς ἐν τῇ νήσῳ μάχης, ἑβδομήκοντα ἡμέραι καὶ δύο.
2 τούτων περὶ εἴκοσιν ἡμέρας, ἐν αἷς οἱ πρέσβεις περὶ τῶν σπονδῶν ἀπῇσαν, ἐσιτοδοτοῦντο, τὰς δὲ ἄλλας τοῖς ἐσπλέουσι λάθρᾳ διετρέφοντο. καὶ ἦν σῖτος ἐν τῇ νήσῳ καὶ ἄλλα βρώματα ἐγκατελήφθη· ὁ γὰρ ἄρχων Ἐπιτάδας ἐνδεεστέρως ἑκάστῳ παρεῖχεν ἢ
3 πρὸς τὴν ἐξουσίαν. οἱ μὲν δὴ Ἀθηναῖοι καὶ οἱ Πελοποννήσιοι ἀνεχώρησαν τῷ στρατῷ ἐκ τῆς Πύλου ἑκάτεροι ἐπ' οἴκου, καὶ τοῦ Κλέωνος καίπερ μανιώδης οὖσα ἡ ὑπόσχεσις ἀπέβη· ἐντὸς γὰρ εἴκοσιν ἡμερῶν ἤγαγε τοὺς ἄνδρας, ὥσπερ ὑπέστη.

XL. Astonishment of Greece at the surrender.

Παρὰ γνώμην τε δὴ μάλιστα τῶν κατὰ τὸν πόλεμον τοῦτο τοῖς Ἕλλησιν ἐγένετο (τοὺς γὰρ Λακεδαιμονίους οὔτε λιμῷ οὔτ' ἀνάγκῃ οὐδεμιᾷ ἠξίουν τὰ ὅπλα παραδοῦναι, ἀλλὰ ἔχοντας καὶ μαχομένους ὡς
2 ἐδύναντο ἀποθνῄσκειν), ἀπιστοῦντές τε μὴ εἶναι τοὺς παραδόντας τοῖς τεθνεῶσιν ὁμοίους, καί τινος ἐρομένου ποτὲ ὕστερον τῶν Ἀθηναίων ξυμμάχων δι' ἀχθηδόνα ἕνα τῶν ἐκ τῆς νήσου αἰχμαλώτων εἰ οἱ τεθνεῶτες αὐτῶν καλοὶ κἀγαθοί, ἀπεκρίνατο αὐτῷ πολλοῦ ἂν ἄξιον εἶναι τὸν ἄτρακτον (λέγων τὸν οἰστὸν), εἰ τοὺς ἀγαθοὺς διεγίγνωσκε, δήλωσιν ποιούμενος ὅτι ὁ ἐντυγχάνων τοῖς τε λίθοις καὶ τοξεύμασι διεφθείρετο.

XLI. The Messenians use Pylus as a base for raids. Alarm of the Spartans, and unsuccessful negotiations for peace.

Κομισθέντων δὲ τῶν ἀνδρῶν οἱ Ἀθηναῖοι ἐβούλευσαν δεσμοῖς μὲν αὐτοὺς φυλάσσειν μέχρι οὗ τι ξυμβῶσιν, ἢν δ' οἱ Πελοποννήσιοι πρὸ τούτου ἐς τὴν γῆν ἐσβάλλωσιν, ἐξαγαγόντες ἀποκτεῖναι. τῆς δὲ 2 Πύλου φυλακὴν κατεστήσαντο, καὶ οἱ ἐκ τῆς Ναυπάκτου Μεσσήνιοι ὡς ἐς πατρίδα ταύτην (ἔστι γὰρ ἡ Πύλος τῆς Μεσσηνίδος ποτὲ οὔσης γῆς) πέμψαντες σφῶν αὐτῶν τοὺς ἐπιτηδειοτάτους ἐλῄζοντό τε τὴν Λακωνικὴν καὶ πλεῖστα ἔβλαπτον ὁμόφωνοι ὄντες. οἱ δὲ Λακεδαιμόνιοι ἀμαθεῖς ὄντες ἐν τῷ πρὶν χρόνῳ 3 λῃστείας καὶ τοιούτου πολέμου, τῶν τε Εἱλώτων αὐτομολούντων καὶ φοβούμενοι μὴ καὶ ἐπὶ μακρότερον σφίσι τι νεωτερισθῇ τῶν κατὰ τὴν χώραν, οὐ ῥᾳδίως ἔφερον, ἀλλὰ καίπερ οὐ βουλόμενοι ἔνδηλοι εἶναι τοῖς Ἀθηναίοις, ἐπρεσβεύοντο παρ' αὐτοὺς καὶ ἐπειρῶντο τήν τε Πύλον καὶ τοὺς ἄνδρας κομίζεσθαι. οἱ δὲ 4 μειζόνων τε ὠρέγοντο καὶ πολλάκις φοιτώντων ἀπράκτους αὐτοὺς ἀπέπεμπον. ταῦτα μὲν τὰ περὶ Πύλον γενόμενα.

NOTES

ABBREVIATIONS

absol.	absolute.	mid.	middle.
accus.	accusative.	neg.	negative.
act.	active.	n. or neut.	neuter.
adj.	adjective.	nom.	nominative.
adv.	adverb.	opt.	optative.
aor.	aorist.	part.	particle.
ch.	chapter.	pass.	passive.
comp.	comparative.	pers.	person.
conj.	conjunction.	pf.	perfect.
contr.	contracted.	pl.	plural.
dat.	dative.	plpf.	pluperfect
dem.	demonstrative.	prep.	preposition.
depend.	dependent.	pres.	present.
f. or fem.	feminine.	pron.	pronoun.
fut.	future.	ptcp.	participle
gen.	genitive.	reflex.	reflexive
intr.	intransitive.	relat.	relative.
impers.	impersonal.	sing.	singular.
impf.	imperfect.	str.	strong.
indic.	indicative.	subjunct.	subjunctive.
inf.	infinitive.	subord.	subordinate.
intr.	intransitive.	superl.	superlative.
lit.	literally.	tr.	transitive.
m. or masc.	masculine.		

(*Numbers refer to sections*)

CHAPTER I

1. **τοῦ δ' ἐπιγιγνομένου θέρους**, i.e., the summer, or rather spring, of 425 B.C., the seventh year of the war between Athens and Sparta.

περὶ σίτου ἐκβολήν, "about the time when the corn puts forth (*ἐκβάλλει*) ears." A similar indication of time is given in ch. II., *πρὶν τὸν σῖτον ἐν ἀκμῇ εἶναι*, "before the corn was ripe."

Λοκρίδες. These were from Locri Epizephyrii, a town in the south-west extremity of Italy, founded by settlers from the Locrians in Greece.

Μεσσήνην. Messana (now Messina) was on the Sicilian side of the straits between Italy and Sicily, nearly opposite Rhegium on the Italian side.

2. **προσβολὴν ἔχον...**, "offering a base of attack on Sicily."

μὴ...ποτε σφίσι μείζονι παρασκευῇ ἐπέλθωσιν. A "larger expedition" was actually sent against Syracuse in 415, but ended in a great disaster for Athens.

3. **ἐς τὴν Ῥηγίνων [γῆν].** Rhegium was an ally of Athens, and therefore likely to attack Messana when it revolted from the Athenians.

ἐπιβοηθῶσι. The verb here practically = "attack," but still retains the original force of *βοηθῶ*, "come to the rescue," in so far as it implies that the Rhegines would be coming to the help of Athens or the pro-Athenian element in Messana against the revolted Messanians (*τοῖς Μεσσηνίοις*).

ἐστασίαζε. In most Greek cities there was a continual feud between the oligarchs or aristocrats and the democratic party, like that between the patricians and the plebeians in Rome. During the war between Athens and Sparta the strife between the two factions became particularly violent everywhere, the oligarchs as a rule taking the side of Sparta and the democrats favouring Athens, as the leading democracy of Greece.

μᾶλλον ἐπετίθεντο. The impf. implies "*were* all the more *for* attacking."

4. **δῃώσαντες**, supply *τὴν γῆν*.

ἄλλαι [αἱ πληρούμεναι] ἔμελλον, "others which were being manned were intended to..." The completion of these preparations is mentioned in ch. XXIV., *τὸ ἄλλο ναυτικὸν ὃ παρεσκευάζοντο προσκομίσαντες*.

CHAPTER II

1. **ὑπὸ δὲ τοὺς αὐτοὺς χρόνους...** In describing the course of the war Thucydides follows the method of the annalists, stringing together all the events that happened during the same year or season, however far apart the places where they occurred, and however little connection there may be between them other than that of time. This leads to such abrupt transitions from Sicily to Greece and back again to Sicily as we find here and in ch. XXIV., and the digression to Thrace in ch. VII.

ἐσέβαλον ἐς τὴν Ἀττικήν. A similar invasion of Attica by the Peloponnesian army, i.e., the Spartans and their allies, took place regularly every year, and was followed with almost equal regularity by reprisals on the part of the Athenian fleet along the coast of the Peloponnese.

2. **τάς τε τεσσαράκοντα ναῦς**, "*the* forty ships." Thucydides has already mentioned these ships in III. 115. They were sent at the request of the allies of Athens in Sicily, and Pythodorus, one of the three commanders, had sailed on ahead with a few of them already.

στρατηγούς. Ten "commanders" were appointed by the Athenians every year, and three of those chosen for this year were sent on the expedition to Sicily.

3. **Κερκυραίων...τῶν ἐν τῇ πόλει**. Corcyra, now Corfu, opposite the coast of Epirus, had been the scene of a violent struggle between the democratic pro-Athenian party and the pro-Spartan oligarchs. This is described at some length by Thucydides (III. 70—85), as an instance of the party conflicts that took place all over Greece during the Peloponnesian war. At Corcyra the democrats gained the victory and massacred all their opponents except five hundred, who escaped to the mainland. These exiles had afterwards returned to the island and occupied Mt. Istone (*τῷ ὄρει*), and were now making effective raids on the lands of the democrats holding the city (*τῶν ἐν τῇ πόλει*).

4. **Δημοσθένει**. Demosthenes, though not always successful, was the most enterprising of the Athenian commanders at this time. In the previous year he had failed disastrously in an attempt to force the Aetolians into an alliance with Athens; on the other hand he had successfully defended the Athenian stronghold of Naupactus when threatened by a combined force of Aetolians and Spartans, and had inflicted a severe defeat on an army of Ambraciots and Spartans which was attacking Acarnania, a country allied with Athens (III. 94—114). After his failure in Aetolia he had not ventured to return to Athens, until his victory in Acarnania restored his reputation. He had not been elected one of the ten *στρατηγοί* for 425, and was therefore now without any official position (*ἰδιώτης*). During the Sicilian expedition (415—413) he was sent in command of reinforcements to Syracuse and made a vigorous effort to restore the fortunes of the Athenian army, but was captured with the rest and put to death by the Syracusans.

αὐτῷ δεηθέντι, "himself having asked," i.e., "at his own request."

CHAPTER III

1. **ἐγένοντο** here = "arrived."

ἠπείγοντο. Note again the impf., "were for hastening."

ἐκέλευε. The impf. gives the idea of failure to persuade; so *ἠξίου* and *ἔπειθεν* below.

πράξαντας ἃ δεῖ... The ptcp. here contains the principal point of the clause. "To sail after having done what was needful" = "to do what was needful before they sailed." The pres. tense *δεῖ* is due to the "graphic construction," which gives in subord. clauses the actual tense and mood used by speakers in the past. Cf. *εἰ μή σφισι πιστεύουσι* in ch. XXVII. 3.

2. **ἐπὶ τοῦτο γὰρ ξυνεκπλεῦσαι.** The inf. is explained by a verb of saying, e.g. *ἔφη*, which must be supplied; "for (he said) it was for this purpose he had sailed with them," i.e., to seize and fortify any useful point along the Peloponnesian coast.

ἀπέφαινε... The verb governs first a simple accus., *πολλὴν εὐπορίαν*, and then an accus. and ptcp. construction, *καρτερὸν ὂν...αὐτό.* "He pointed out abundance of timber (how abundant the timber was)...and that it (the place) was strong."

ἐπὶ πολύ must be translated here simply as = *πολύ*; what was originally an adverbial phrase, "over a wide extent," has become practically equivalent to a noun, "a wide extent." *τῆς χώρας*, i.e., the country round.

σταδίους...τετρακοσίους, a little less than 46 miles. Pylus is on the west coast of Messenia; its harbour is now known as the bay of Navarino, where the British, French and Russian fleets utterly defeated the Turks in 1827 and brought about the liberation of Greece.

ἐν τῇ Μεσσηνίᾳ ποτὲ οὔσῃ γῇ, "in the territory which was once Messenian." Messenia, the south-west province of the Peloponnese, had been finally conquered by the Spartans in 460, and the inhabitants had been either driven into exile or reduced to a state of serfdom. They remained bitter enemies of the Spartans and were most useful allies of Athens in this war.

Κορυφάσιον, from *κορυφή*, "head, summit."

3. **δαπανᾶν**, usually "to spend," here "to put to expense."

λιμένος τε προσόντος. καὶ τοὺς Μεσσηνίους...πλεῖστ' ἂν βλάπτειν. The construction here is irregular, as is often the case with participles in Thucydides. After the gen. absol. *λιμένος προσόντος*, "since there was a harbour," a verb of thinking, which should properly be in the ptcp. and agree with *τῷ δέ*, must be understood from *ἐδόκει* above to account for the accus. and inf. *καὶ τοὺς Μεσσηνίους...πλεῖστ' ἂν βλάπτειν*, "and (since he thought that) the Messenians would do much harm."

ὁμοφώνους τοῖς Λ., "speaking the same dialect as the Lacedaemonians," i.e., Doric, the dialect of Greek used in the Peloponnese generally.

CHAPTER IV

1. **τοῖς ταξιάρχοις.** These were ten Athenian officers, subordinate to the *στρατηγοί*, each commanding the infantry of his own tribe.

2. **λογάδην...φέροντες λίθους**, "bringing stones as they picked them out."

ὡς ἕκαστόν τι ξυμβαίνοι, "according as each fitted in (with the rest)." The optative is used with *εἰ* or a relat. pron. or adv. to express anything repeated in an indefinite number of instances in the past. Cf. *εἴ που δέοι* below, *εἴ που καὶ δοκοίη* in ch. xi., and *ὁπότε ἄνεμος εἴη* in ch. xxiii.

καὶ τὸν πηλόν... This very precise account of the way in which the soldiers carried the mortar is one of the picturesque touches introduced by Thucydides into his narrative every now and then. The most famous is his description of the Athenian army watching from the shore the sea-fight in the Syracusan harbour, and the various cries and gestures with which they followed the changing fortunes of their fleet.

εἴ που δέοι. *εἰ* with indefinite prons. or advs. must be translated "any...that," or "whoever," "wherever" etc. *εἴ που*, "at any place where" or "wherever."

ὡς μάλιστα μέλλοι, "in the way in which it was most likely to."

ἐπιμένειν. The fut. is more common than the pres. inf. after *μέλλω*.

3. **φθῆναι...ἐξεργασάμενοι.** *φθάνω* may be used, as here, with another verb in the ptcp., "to be beforehand...having completed," i.e., "to complete...before..." But it may also be used in the ptcp. itself, and translated as an adv. belonging to the principal verb, as in ch. viii., *ὑπεκπέμπει φθάσας*, "he sends out secretly... having been beforehand," i.e., "sends out beforehand."

CHAPTER V

1. **ἑορτήν τινα.** Religious festivals were always strictly observed by the Spartans, and would prevent them from taking any military action, however necessary. Another instance of this occurred in 479, when the Spartans refused to send help to the Athenians against the Persian invaders of Attica, because they happened to be celebrating the festival of the Hyacinthia (Herodotus, IX. 7). The same kind of scrupulousness had prevented them from aiding the Athenians at the battle of Marathon, during the first Persian invasion in 490; the Athenian messenger arrived on the ninth day of the month, but the Spartans were forbidden by law from marching on that and the following days, until the full moon (Herod. VI. 106).

ὡς...οὐχ ὑπομενοῦντας σφᾶς, "in the belief that they would not await them," *σφᾶς*, i.e., their attack. *ὡς* with the ptcp. must often be translated "in the belief that" or "on the assumption that." *οὐχ ὑπομενοῦντας*, with which *αὐτοὺς* must be supplied, is probably an instance of the accus. absol., though this is rare except with impers. verbs. It might, however, be taken as object of *ἐν ὀλιγωρίᾳ ἐποιοῦντο*, "they held them in contempt...as not likely to await them." *ληψόμενοι* is nom. as referring to the subject of *ἐποιοῦντο*, "as being (themselves) likely to take (the place)," i.e., "in the belief that they would take..."

ὁ στρατὸς ἔτι ἐν ταῖς Ἀθήναις ὤν, "the fact that their army was still before (*ἐν*) Athens."

CHAPTER VI

1. **ἐπύθοντο τῆς Πύλου κατειλημμένης,** "learned that Pylus had been occupied." *πυνθάνομαι* may be followed either by the accus. or gen. with the participle.

νομίζοντες μὲν οἱ Λακεδαιμόνιοι. The original subject of the sentence, *οἱ ἐν τῇ Ἀττικῇ ὄντες Πελοποννήσιοι*, is here limited to the Spartans, because they alone were affected by the first reason for retreat, the thought "that the matter of Pylus touched them nearly" (*οἰκεῖον σφίσι τὸ περὶ τὴν Πύλον*).

τοῖς πολλοῖς, either "for the greater number," or "for their large numbers."

μείζων παρὰ τὴν καθεστηκυῖαν ὥραν, "greater (i.e. worse) than was usual at that season," lit. "greater contrary to that season."

The comparative in Greek may mean, as here, "more than usual," and the prep. *παρά* is due to the idea that the weather was "contrary to (the rule at) that season." *τὴν καθεστηκυῖαν*, lit. "that had set in." *καθεστηκώς*, the pf. ptcp. of *καθίστημι*, may often be translated "the present," or "the then (existing)."

2. **ξυνέβη** is followed by accus. and inf., "it came to pass that."

CHAPTER VII

1. **Ἠϊόνα τὴν ἐπὶ Θρᾴκης.** The site of this Eion "in the Thraceward region" is uncertain, but it was probably in the peninsula of Chalcidice. It must be distinguished from the town of the same name at the mouth of the river Strymon. This latter Eion was saved for the Athenians two years later by Thucydides himself, then in command of an Athenian fleet, though he failed to save the more important town of Amphipolis three miles away, and was banished from Athens for his failure.

Μενδαίων. Mende was a town situated on Pallene, the furthest west of the three promontories of Chalcidice.

προδιδομένην. The pres. tense implies that the surrender was *offered* to him by traitors. The word might be used of a town that was never actually taken, and points here to the fact that it was not held.

Χαλκιδέων, inhabitants of Chalcidice, the peninsula south-east of Salonica, between what are now called the gulfs of Salonica (Saloniki) and Orfano.

Βοττιαίων. This tribe had been driven southwards from its original home by the Macedonians, and had settled in Chalcidice. It is associated with the Chalcidians in I. 65 and II. 79.

CHAPTER VIII

1. **περιοίκων**, "provincials," i.e., the inhabitants of Laconia outside Sparta. They occupied a position somewhat subordinate to the citizens of Sparta itself, the *Σπαρτιᾶται*, and did not enjoy the same political rights. The word *Λακεδαιμόνιοι* includes both *περίοικοι* and *Σπαρτιᾶται*.

2. **περιήγγελλον.** This verb may mean "send an order" as well as "send a message," and as a verb of commanding it is followed here by the inf. *βοηθεῖν*.

ὑπερενεχθεῖσαι, "conveyed across" by means of hauling-engines (*ὁλκοί*). Ships were conveyed overland in this way on many occasions. Cf. III. 81, VIII. 8, and Livy XLII. 16, per Isthmi iugum navibus traductis. The Greek trireme was large enough to hold a crew of 200.

τὸν Λευκαδίων ἰσθμόν. Leucas or Leucadia, now Santa Maura, an island off the coast of Acarnania, was originally joined to the mainland by a sandy isthmus. The Corinthians dug a canal here, but it afterwards became filled up with sand. Leucadia is now separated from the mainland by a lagoon varying in width from 100 yards to a mile and a half.

λαθοῦσαι. The naval reputation of the Athenians was so great that the Spartans, even when superior in numbers, as on this occasion, never showed any anxiety to meet their enemies at sea.

τὰς ἐν Ζακύνθῳ Ἀττικὰς ναῦς, the Athenian fleet on its way to Corcyra. Zacynthus, where it had now arrived, is an island off the north-western coast of the Peloponnese, now called Zante.

3. **ἀγγεῖλαι**, the inf. of purpose used after a verb of motion (*ὑπεκπέμπει*).

4. **οἰκοδόμημα** is a somewhat contemptuous word; it might be translated "an erection" or "a fabric."

ἀνθρώπων ὀλίγων ἐνόντων, "with only a few men in it." Thucydides begins with a ptcp. *εἰργασμένον* agreeing with *οἰκοδόμημα* and joins on to it with a *καί* a gen. absol. qualifying the same noun.

5. **ὅπως μὴ ᾖ**, "in order that it might not be possible." *ᾖ* is the subjunct. of the impers. *ἔστι*. The subjunct., the more graphic mood, is often substituted for the opt. after a historic tense; cf. *ἧν* ...*ἕλωσι* above.

ἐφορμίσασθαι ἐς αὐτόν, "come to anchor in it." The prep. *ἐς* is due to the idea of "motion to" implied in *ἐφορμίσασθαι*.

6. **ἡ γὰρ νῆσος ἡ Σφακτηρία καλουμένη**. Sphacteria, now called Sphagia, stretches from north to south, and is 4,800 yards long, not 3000 yards (*πέντε καὶ δέκα σταδίους*), the length given by Thucydides. Between its northern extremity and Pylus, which lies opposite, is the first of the two entrances to the harbour (the bay of Navarino); the width is 130 yards. The other entrance is between the southern end of Sphacteria and the mainland to the east, where the modern Navarino lies; it has a width of 1300 yards, enough for twenty triremes, not eight or nine, if we keep to

the same proportion as that for the northern entrance. It would be impossible to block so wide a space with ships "having their prows towards the enemy" (ἀντιπρῴροις), i.e., facing outwards. These difficulties may be explained by assuming either that Thucydides was misled by his authorities, or that the physical features of the bay have changed since his time.

ἐγγὺς ἐπικειμένη, "lying near to the mainland."

μέγεθος, "in length," accus. of respect.

7. **τὴν δὲ νῆσον ταύτην...**, "and fearing (as regards) this island lest they (the Athenians) should make war upon them from it," i.e., "fearing lest the A. should use this island as a base for making war upon them."

8. **οὕτω γὰρ...τήν τε νῆσον πολεμίαν ἔσεσθαι.** The infs. in this sentence (*ἔσεσθαι, ἕξειν, ἐκπολιορκήσειν*) are due to the fact that these were the *thoughts* of the Spartans; a verb of thinking or hoping must be understood. "For so they thought that .."

πρὸς τὸ πέλαγος, "facing the open sea," i.e., on the western side.

σίτου τε οὐκ ἐνόντος καὶ...κατειλημμένον, a similar construction to *εἰργασμένον καὶ ἀνθρώπων ὀλίγων ἐνόντων* above, but with the positions of the gen. absol. and the ptcp. agreeing with the object reversed. Translate, "since there was no food in it, and it had been built in haste," lit. "with little preparation."

9. **καὶ διεβίβαζον.** *καί* expresses the idea that they not only thought but acted. The impf. tense implies that they sent several detachments in turn.

Εἵλωτες οἱ περὶ αὐτούς. The Helots were the serfs of the Spartans, some of them descended from the original inhabitants of Laconia, who were enslaved by the Dorian invaders, and others from the Messenians who did not leave their country after the Spartan conquest. They accompanied their masters to war, generally as light-armed troops, and might be set at liberty if they distinguished themselves.

CHAPTER IX

1. **προσβάλλειν.** The pres. inf. is here used after *μέλλω* instead of the more regular future tense.

αἳ περιῆσαν. Two of the five ships left with Demosthenes at Pylus had been sent to warn Eurymedon and Sophocles.

προσεσταύρωσε, "joined them to the fortification by a palisade." *προεσταύρωσεν*, "raised a palisade in front of them," has been suggested as a better reading.

2. **ἀόπλων,** i.e. without *heavy* armour (*ὅπλα*).

πρὸς τὴν ἤπειρον, "on the side towards the mainland," i.e., to the north.

ἐκ πάντων. It has been estimated that the whole Athenian force under Demosthenes only amounted to 640 men.

σφίσι includes the other Athenians with Demosthenes, "since *their* fortification."

ἐσβιάσασθαι..., "he thought that they would be eager to force the approaches here." The reading of the MSS. is *ἐπισπάσασθαι*, which is too full of difficulties to stand. *ἐπισπᾶσθαι* is used in the sense of "induce," and the meaning of the sentence, with the reading of the MSS., would have to be "he thought that he would induce them to be eager" to attack at this point. But the fut. inf. *ἐπισπάσεσθαι*, not the aor., would be required, while on the other hand there would be no reason why *προθυμήσεσθαι* should be in the future tense.

3. **ἐκείνοις,** with *ἁλώσιμον*, "easy for them to take."

γίγνεσθαι, inf. depending on a verb of thinking which must be understood from *ἡγεῖτο* above. The verb *γίγνεσθαι*, though pres. inf., points to the future, like the English word "become." Translate, "he thought that the place became easy for them to take if they forced a landing."

4. **πρὸς αὐτὴν τὴν θάλασσαν,** "right down to the sea."

ὡς εἴρξων, understand *τὴν ἀπόβασιν*, "with the intention of preventing the landing."

τοιάδε. The words *τοιόσδε* and *τοσόσδε* refer to what follows, *τοιοῦτος* and *τοσοῦτος* to what has preceded. Cf. *τοσαῦτα* in ch. XI. Translate, "as follows."

CHAPTER X

In the speeches he introduces into his narrative Thucydides does not pretend to give anything like the actual words of the speaker, but only what might probably have been said in the circumstances. They serve to explain and comment on the motives and actions of statesmen or nations, as the choric odes in a Greek tragedy comment on the motives and actions of the leading characters. They are too elaborately constructed in style and too full of generalizations to have been addressed even to such a quick-witted audience as the Athenians. It is always Thucydides who is speaking; he never changes his style to suit the character of the man into whose mouth the words are put. Even the speech here attributed to Demosthenes, though simple and straightforward compared with some others, contains sentences too intricate to have been addressed by a soldier to soldiers.

The speech may be briefly summarised as follows: "You must meet the enemy in a spirit of confidence, without considering the danger too closely. If only you stand firm you will have several advantages over the enemy, for the difficulties of the ground will then be in your favour, and the Spartans, though more numerous, can only attack in small detachments at a time. Besides, they will have to fight from ships, a much harder task than to fight on even terms by land. As Athenians and experienced sailors you must be well aware how difficult it is to force a landing if the men on shore stand their ground, and the remembrance of this should encourage you to meet the enemy's attack more than half-way. Go right down to the water's edge and make that your first line of defence."

1. **μᾶλλον δ' ἀπερισκέπτως εὔελπις...**, "but rather go to meet the enemy without pausing for reflection, confident that you will come out safely even from these dangers." The ptcp. *περιγενόμενος* with *ἂν* must depend on *εὔελπις*, but this construction, instead of the usual fut. inf. after verbs of hoping, is difficult to explain. To insert *ὡς* after *τοῖς ἐναντίοις*, where it might easily have dropped out in the MSS., would make the sentence quite regular; *ὡς... περιγενόμενος*, "in the belief that you will come out safely (in that case)."

κινδύνου...προσδεῖται, "need the speediest putting of the matter to the hazard," i.e., "require that the danger should be faced at once." *κίνδυνος* is here used for "facing the danger."

2. **τὰ ὑπάρχοντα ἡμῖν κρείσσω**, "the advantages we already possess."

3. **ἡμέτερον**, "ours," i.e., "in our favour."

ὑποχωρήσασι δέ..., lit. "but for us having retreated it will be easy (for our enemies) with no one to hinder them," i.e., "if we retreat it will be easy." In the first clause Thucydides used the gen. absol. *μενόντων ἡμῶν*, "if we stand firm," but in this he changes the construction to the "dat. of advantage or disadvantage," *ὑποχωρήσασι δὲ* (*ἡμῖν*), with the ptcp. again used as conditional.

βιάζηται, used in a pass. sense, like *βιάζοιτο* below, "if he is forced back."

ἐπὶ γὰρ ταῖς ναυσὶ.... The connection of this parenthesis with the preceding words is not clear, except that it adds another reason for expecting a desperate resistance from the enemy if driven back after landing.

4. **καὶ οὐκ ἐν γῇ στρατός ἐστιν**, "and it (the Spartan force) is not a larger force fighting on equal terms with us by land, but one fighting from ships, which require the meeting of many fortunate contingencies at sea." To emphasize the *ξύν* in *ξυμβαίνειν*, so as to make it mean "occur at the same time," gives more point to the sentence, but the word might mean simply "occur."

5. **ἀντιπάλους (εἶναι)**, "are enough to compensate for."

πλήθει, here "*inferiority* of numbers."

ἐπισταμένους...τὴν ναυτικὴν...απόβασιν, "knowing by experience *as regards* landing from ships." The construction is the same as that of *τὴν νῆσον ταύτην φοβούμενοι* in ch. VIII.

ἐμπειρίᾳ. Raids were often made from the Athenian fleets on the coast of the Peloponnese, in retaliation for the Spartan invasions of Attica.

βιάζοιτο. The subject may be either *τις* or *ἡ ἀπόβασις*.

καὶ αὐτοὺς νῦν μεῖναι, "(I call upon) you too to stand firm," like the men on whose territory they had landed themselves.

τὴν ῥαχίαν, "the breakers," or "the rocks on the shore."

CHAPTER XI

2. **τεσσαράκοντα καὶ τρισί.** We are not told what had become of the other seventeen out of the sixty ships that sailed to Corcyra.

3. **ἔκ τε γῆς καὶ ἐκ θαλάσσης**, "landwards and seawards."

διελόμενοι, "dividing (their fleet)."

εἴ πως...ἕλοιεν, "in the hope that by some means they might take the fort." *εἰ* with the opt., like *si* with the subjunct. in Latin, is sometimes used in the sense "if perchance (they) might," i.e., "in the hope that they might."

4. **Βρασίδας.** This commander was the bravest and most

energetic Spartan of his time. In II. 69, 79 he is mentioned as taking part in an expedition to Corcyra, and as trying unsuccessfully to inspire his fellow-commander Alcidas with his own vigour. A little later he and other Spartan commanders planned an attack on the Piraeus, the harbour of Athens, but this was not carried out owing to fear of the risk, a fear we cannot imagine Brasidas himself to have shared. In 424 he went on an expedition to the north and won over from the Athenians several towns in Chalcidice and the neighbouring region, of which the most important was Amphipolis. His success was chiefly due to his personal popularity, for he possessed the faculty, not common among his countrymen, of conciliating other states and winning their trust. During the following year he accompanied Perdiccas, the king of Macedonia, then allied with Sparta, in an unsuccessful campaign against the Lyncestae, a Macedonian tribe which had rebelled. In 422 he won a victory against the Athenians near Amphipolis, but was killed in the battle, as was also the Athenian commander Cleon.

εἴ πῃ καὶ δοκοίη..., "wherever it did seem possible." *δοκοίη* is the indefinite opt. of repeated action in past time, after *εἰ* or a relat. pron. or adverb.

ξύλων φειδομένους, "sparing timber." The word *ξύλων* expresses the contempt of a military man for ships.

τοὺς πολεμίους...περιιδεῖν...πεποιημένους, "to leave their enemies undisturbed when they had built a fortress in the land." *περιορᾶν* may be followed by an accus. and ptcp., e.g. *περιορᾶν τινα τοῦτο ποιοῦντα*, "to overlook someone doing this," or by an accus. and inf., e.g. *περιορᾶν τινα τοῦτο ποιεῖν*, "to allow someone to do this."

τοῖς Λακεδαιμονίοις...ἐπιδοῦναι, "to sacrifice for the Lacedaemonians."

παντὶ τρόπῳ, "in any way."

CHAPTER XII

1. **τραυματισθεὶς πολλά**, "having received many wounds." The neut. pl. *πολλά* is here used adverbially.

παρεξειρεσίαν, the space at either end of the ship where there were no rowers' benches. Translate, "the bows of the ship."

τροπαῖον (from **τροπή**, "rout"). After winning a battle Greek armies hung up on a post fixed in the ground some of the weapons and armour taken from the enemy, as a sign of victory.

τῆς προσβολῆς ταύτης, "for (the defeat of) this attack."

2. **χαλεπότητι καὶ τῶν Ἀθηναίων μενόντων**, "owing to the difficulty...and because the A. stood firm." The reasons are given in two different constructions, the dat. of the instrument and the gen. absol., connected by a *καί*.

3. **ἐς τὴν ἑαυτῶν (γῆν) πολεμίαν οὖσαν**, "on their own coast, which had now become hostile ground."

ἐπὶ πολὺ γὰρ ἐποίει..., lit. "for to be landsmen and strongest in military matters brought (made) to the one side (the Spartans) much of their reputation, while to be seamen and have the advantages in ships (did the same) for the others," i.e. the Athenians. Translate:—"The reputation of the Spartans was largely due to their being landsmen..., and that of the Athenians to their seamanship and naval superiority." The subjects of *ἐποίει* are the infins. *εἶναι* and *προὔχειν*; it is as though Thucydides had written *τὸ ἠπειρώτας εἶναι* ("the being landsmen")...*ἐπὶ πολὺ ἐποίει*, but the accus. *ἠπειρώτας* is attracted into the dat. case from *τοῖς μέν*, the indirect object of *ἐποίει*. *ἐπὶ πολύ* = *πολύ*, as in ch. III, and is the direct object of *ἐποίει*.

CHAPTER XIII

1. **ἐπέπαυντο**, "they had now ceased"; the plpf. prepares us for a new move on the part of the Spartans.

ἐλπίζοντες..., "hoping that, though the wall on the side of the harbour was high (had height), they could take it...." The *μέν* clause is not, of course, part of what they hoped, though the accus. and inf. construction is due to the verb *ἐλπίζοντες*; it is equivalent, as such clauses often are, to a gen. absol. preceded by *καίπερ*, and must be begun with "while" or "though" in translating. *ἀποβάσεως μάλιστα οὔσης*, "since a landing was easiest (lit. "most") there." *ἄν* after *ἑλεῖν* is not necessary, as the aor. inf. is sometimes used instead of the fut. after *ἐλπίζω*.

2. **τεσσαράκοντα**. Some of the original forty ships had been sent to Sicily, and five of them had been left with Demosthenes, though two of these had returned to the fleet to warn the Athenian commanders of the attack on Pylus. The next sentence explains how the number was made up again.

Ναυπάκτου. Naupactus, now Lepanto, a town of the Locri Ozolae on the northern coast of the gulf of Corinth, was originally occupied by the Athenians in 460 and handed over in the next year to Messenian settlers who had been driven out of their own country by the Spartans.

Χῖαι. Chios was one of the most important islands subject to Athens, and possessed a powerful fleet. It revolted in 412, after the disaster to the Athenians in Sicily.

3. **καθορμίσωνται**, deliberative subjunct., "where they were to come to anchor."

Πρωτήν. This island, now called Prodano, is eight miles to the north-west of Pylus.

ἦν... ἐθέλωσι, "in case they should be willing."

4. **οὔτε ἃ διενοήθησαν...** If they had really intended to block the entrances their neglect to do so would be a characteristic example of Spartan dilatoriness. But the southern entrance, if as wide then as it is now, could not have been blocked by triremes in the way Thucydides describes (ch. VIII), so that the Spartans could not have had the intention Thucydides attributes to them.

ἐπλήρουν, "*began* to man."

CHAPTER XIV

1. **τὰς μὲν πλείους**, accus. governed by *ἐς φυγὴν κατέστησαν*, "put to flight," not by *προσπεσόντες*, which takes a dative.

ὡς διὰ βραχέος, "as well as they could for such a short distance," the Spartan ships being near the land. *μετεώρους* above cannot mean "out in the open sea," for the Spartan ships were evidently still in the harbour.

αὐτοῖς ἀνδράσι, "men and all," "with the crew."

ταῖς δὲ λοιπαῖς, i.e., the rest of the ships that had been manned.

2. **ἀπελαμβάνοντο**, "*were being* cut off," by the defeat of the Spartan fleet, which was still going on before the eyes of the men on shore.

ἐπεσβαίνοντες..., "entering the sea even in full armour to meet the enemy" (*ἐπί*).

ἀνθεῖλκον..., another graphic touch. The rescue of a fleet by an army was not an uncommon event. In 413 the Athenian soldiers saved some of their ships from the victorious Syracusans in the harbour of Syracuse, and at the battle of Abydus in 411 a Persian army saved the greater part of a Peloponnesian fleet which was being towed away by the Athenians.

κεκωλῦσθαι. The subject of this verb is *ἔργον*, which instead of being made the antecedent to *ᾧ...τινι* has been put inside the relat. clause and attracted into the case of the relative. Translate :—"Each man thought that any fighting (action) in which he himself did not take part was going the worse for his absence." *κεκωλῦσθαι*, lit. "had been hindered," i.e., by his absence.

3. **ἀντηλλαγμένος** properly agrees with *ὁ θόρυβος*, but in translating some word like "fighting" or "battle" must be introduced.

οἵ τε γὰρ Λακεδαιμόνιοι..., a somewhat similar antithesis to that in ch. XII. 3.

ὡς εἰπεῖν, "one may say," "practically."

ἄλλο οὐδὲν ἤ... The word *ἐποίουν* must be understood after **ἄλλο** *οὐδέν*, "were doing nothing but," "were actually" or "in fact."

ὡς ἐπὶ πλεῖστον ἐπεξελθεῖν, "to follow up as far as possible." *ὡς* goes with *πλεῖστον*.

5. **καταστάντες...ἐς**, which may be translated simply as "going to" or "returning to," implies in full the meaning "going to their quarters and settling down there."

ὡς τῶν ἀνδρῶν ἀπειλημμένων, "in the knowledge that the men were cut off."

οἱ δ' ἐν τῇ ἠπείρῳ...βεβοηθηκότες, "the Peloponnesians who were on the mainland having now hurried to the spot from all quarters."

CHAPTER XV

1. **ἔδοξεν**, impers., followed by accus. and inf., "it was resolved that..."

τὰ τέλη, though neut., has a masc. sense ("the magistrates"), and is therefore followed by the masc. pl. of the ptcp., *καταβάντας*. These magistrates were the ephors ("overseers"), who had almost as much power as the two kings of Sparta themselves.

2. **ἀδύνατον ὄν**, the ptcp. construction, here impers., used after verbs of perceiving.

κινδυνεύειν, followed by the accus. and inf. construction *παθεῖν αὐτούς*, "to run the risk of their suffering any harm."

ποιησαμένους. This should properly be in the dat., agreeing with *αὐτοῖς*, but after an impers. construction beginning with the dat. a change to the accus. is quite common.

CHAPTER XVI

1. **τοιαίδε**, "on the following terms."

Λακεδαιμονίους...παραδοῦναι, "the Spartans were to give up." *ἐγίγνοντο σπονδαὶ τοιαίδε* is followed by accus. and inf. as though *ἐφ' ᾧ* were understood.

σῖτον...τακτὸν καὶ μεμαγμένον, "a fixed amount of dough ready kneaded."

χοίνικας. A *χοῖνιξ* was about a quart and equal to four *κοτύλαι*. A single *χοῖνιξ* would be a slave's allowance; in Herod. VII. 184 it is given as rather an under-estimate of what a Persian soldier might be expected to eat in a day.

δύο κοτύλας = a pint.

ὅσα μὴ ἀποβαίνοντας, "as much as (they could) without landing," i.e., "save that they were not to land."

2. **ὅ τι δ' ἂν...καὶ ὁτιοῦν**, "whatever part of these articles, even the smallest."

λελύσθαι, pf., "were to be regarded as broken." So *ἐσπεῖσθαι* below.

μέχρι οὗ ἐπανέλθωσιν. The regular construction requires *ἄν* with the subjunct. after a relat. pron. or adv., but this is sometimes omitted. Cf. *οὗ βραχεῖς ἀρκῶσι* in ch. XVII. 2.

ἐλθόντων δέ, understand *τῶν πρέσβεων*.

οἵασπερ ἄν, "in just the same state in which they received them."

3. **ἑξήκοντα**, i.e., more than the Athenian fleet at Pylus. Their surrender without a fight shows how far the Spartans were paralysed by the naval reputation of the Athenians. Only forty-three ships had taken part in the sea-fight, and five of these had been taken, but the terms of the armistice included *τὰς ἐν τῇ Λακωνικῇ πάσας*, all those on the Laconian coast.

CHAPTER XVII

Summary of the speech:—"We have been sent to negotiate for the recovery of the men on Sphacteria, and shall speak at some length, as Spartans do when circumstances make it advisable. You have the opportunity now of making peace on favourable terms; do not be carried away by your present success, but act like men who know by experience the uncertainty of fortune, of which our own humiliation may remind you. The wisest men are those who are prepared for ill fortune as for good, who recognize that war may lead them in unexpected directions and are therefore willing to make peace in the hour of success. It will be better for both sides not to fight it out to the end at Sphacteria. We offer you our friendship and alliance, if you on your side will offer us reasonable terms. Remember that generosity is the surest way of turning an enemy into a firm friend, when harshness might drive him to a desperate resistance. If you refuse to make peace, any disaster you may inflict on us will only make us more bitter in our enmity, and the Greeks will hold you responsible for the continuation of the war. As it is you have the opportunity of granting us your friendship as a favour, and as friends our two states will control all Greece."

1. **πράξοντας...**, "to arrange such terms as we may persuade you to accept in your own interests (lit. "as being advantageous to you") and as are at the same time (*τὸ αὐτό*) most likely..." *ὅ τι* is the object of *πείθωμεν* and the subject of *μέλλῃ*. *τὸ αὐτό* should be taken with the second clause in translating.

2. **οὐ παρὰ τὸ εἰωθός**, "not contrary to custom," contains the main point of the sentence. "In speaking at some length we shall not be breaking our custom."

ἐπιχώριον ὄν, accus. absol., "it being the habit of our country," "(acting) according to the habit of our country."

τι τῶν προὔργου, "something important" or "useful."

3. **ὡς ἀξύνετοι διδασκόμενοι**, "with the idea that you are being admonished as though you were unintelligent." *ὡς* goes both with *ἀξύνετοι* and *διδασκόμενοι*.

τοῦ καλῶς βουλεύσασθαι, "(a reminder) *to* take good counsel."

πρὸς εἰδότας, "addressed to men of experience."

4. **οἱ...λαμβάνοντες τῶν ἀνθρώπων**, "those (of) men who receive." *τῶν ἀνθρώπων* is a partitive genitive.

τοῦ πλέονος..., "in their hopes they reach out after more," i.e., "they become ambitious of more."

τὰ παρόντα..., accus. of respect, "at the present time."

5. **ἐπ' ἀμφότερα**, "in both directions," i.e., "for the better and for the worse."

δίκαιοί εἰσι...εἶναι, "are bound to be," "must be," a common use of *δίκαιος* with the infinitive.

ὃ...προσείη, "this feeling would most naturally be implanted in..."

CHAPTER XVIII

1. **οἵτινες.** The antecedent is the *ἡμῶν* implied in *ἡμετέρας*. *ὅστις* often begins a clause explaining what has preceded; translate, "seeing that we..."

αὐτοὶ κυριώτεροι νομίζοντες εἶναι, "thinking that *we* had the more power," "that it lay rather with us." *κύριος εἶναι*, "to have the deciding voice."

2. **αὐτό,** a vague neuter, referring to *ξυμφόρας* above.

οὔτε...ὑβρίσαντες, "nor because we grew insolent owing to the gaining of too great a power." *μείζονος (δυνάμεως) προσγενομένης*, lit. "a greater power having been added to us."

ἀπὸ δὲ τῶν ἀεὶ ὑπαρχόντων, "through a confidence based on the forces we always possessed," lit. "from the forces..."

ἐν ᾧ...ὑπάρχει, "a matter in which the same mistake is possible to all alike."

3. **καὶ τὸ τῆς τύχης,** "the element of fortune also." *τὸ τῆς τύχης* is practically equivalent to *ἡ τύχη*.

4. **σωφρόνων δὲ ἀνδρῶν οἵτινες...** The gen. may be partitive, "they are (among) wise men who...," or it may be due to a fusion of the two constructions *σωφρόνων ἀνδρῶν ἐστι* with inf., "it is (a mark) of wise men to...," and *σώφρονες ἄνδρες εἰσὶν οἵτινες...*, "they are wise men who..."

ἐς ἀμφίβολον...ἔθεντο, "set their prosperity on a sure foundation (*ἀσφαλῶς*) to face good fortune or ill." *ἐς ἀμφίβολον*, "facing both ways," as though to meet an attack on both sides. *ἔθεντο* is the gnomic aor. used to express a general rule; it gives the idea that those who did so in the past were wise, and consequently those who do so in the future will be no less wise.

νομίσωσι. Thucydides here passes to the usual construction after indefinite relatives, but omits the regular *ἄν*, as with *μέχρι οὗ ἐπανέλθωσιν* in ch. XVI.

μὴ καθ' ὅσον...ἡγήσωνται, "that the war does not come home to (*ξυνεῖναι*, lit. "associate with") a man only in so far as he wishes to take part in it (*καθ' ὅσον...μέρος*, lit. "in proportion to as large a part as") but according as their fortunes direct." *αὐτῶν* should properly be *αὐτοῦ*, referring back to *τις*, but Thucydides is thinking of the pl. subject of the whole clause, *οἵτινες*.

τῷ ὀρθουμένῳ αὐτοῦ, "to their success in it," i.e., the war. *τὸ ὀρθούμενον*, lit. "that which is done aright" or "goes right." *αὐτοῦ* is a partitive genitive.

5. **ἃ πολλὰ ἐνδέχεται.** The relat. is used as though *σφαλῆτε* were preceded by a cognate accus. *τῶν σφαλμάτων τι*, "meet with one of the calamities which are possible in abundance" (*πολλά*), i.e., "meet with one of the many calamities that are possible."

ἐξόν, accus. absol. from *ἔξεστι*, "when it is possible."

CHAPTER XIX

1. **διδόντες**, "offering."

ἐς ἀλλήλους ὑπάρχειν, epexegetic inf., "to be (established) between us."

ἐκ τῆς νήσου. We should say "the men *on* the island"; the *ἐκ* is due to the verb *ἀνταιτοῦντες*, "asking for the men (to be given us) from the island."

εἴτε...διαφύγοιεν, depending on impers. *διακινδυνεύεσθαι*, "that the issue should not be fought out to the uttermost, whether they shall escape..." What is called the deliberative subjunct. or opt. (e g. *διαφύγοιεν*) may express deliberation or doubt or the deciding between two alternatives. The subjunct. would be more regular here, after the primary tenses *προκαλοῦνται...ἡγούμενοι.* But the ambassadors are here giving the thoughts of the men who sent them—they say "the Spartans," not "we" as in the next sentence—and as these thoughts belong to the past the verbs are virtually historic.

μᾶλλον ἂν χειρωθεῖεν. The *ἄν* by introducing a conditional element is intended to make this second alternative seem more remote.

2. **κατ' ἀνάγκην ὅρκοις ἐγκαταλαμβάνων**, "binding (his enemy) by oaths under compulsion," i.e., by the oaths taken to confirm the treaty of peace.

μὴ ἀπὸ τοῦ ἴσου ξυμβῇ, "makes peace on unequal terms."

παρόν, from impers. *πάρεστι*, accus. absol. like *ἐξόν* in ch. XVIII. 5.

πρὸς τὸ ἐπιεικές, "leaning to the side of reasonableness," the same use of the prep. as in *πρὸς ὀργήν*.

ἀρετῇ αὐτὸν νικήσας, "surpassing him in generosity."

3. **ἀνταποδοῦναι ἀρετήν**, "to show (lit. "give") generosity in return."

αἰσχύνῃ...ξυνέθετο, "through a sense of honour to abide by the terms to which he agreed." *οἷς = ἐκείνοις ἅ.*

4. **τὰ μέτρια διενεχθέντας**, "divided from them by minor differences." *τὰ μέτρια*, extension of cognate accusative.

πεφύκασι...ἡδονῆς, "and they are by nature inclined to give way gladly in their turn to those who have voluntarily made concessions to them." *ἐνδοῦσιν*, lit. "having given in."

τὰ ὑπεραυχοῦντα = *τοὺς ὑπεραυχοῦντας*.

παρὰ γνώμην διακινδυνεύειν, "to fight it out to the end with an unexpected stubbornness." *παρὰ γνώμην*, "contrary to (general) expectation."

CHAPTER XX

1. **διὰ μέσου γενόμενον,** "occurring in the interval" before peace is made.

ἐν ᾧ...ἔχειν, "in which case we must have against you an undying enmity of our own, besides that we have in common (with our allies)."

2. **ὄντων ἀκρίτων,** understand *τῶν πραγμάτων*.

προσγιγνομένης...κατατιθεμένης. The pres. ptcps. refer to the possibilities if peace is made; "when you may win...and we may put an end to our unfortunate position," lit., "when...may be added to you, and for us our unfortunate position may be ended."

πολεμοῦνται from *πολεμῶ* [-όω], "they are made enemies" or "involved in war."

ἀσαφῶς ὁποτέρων ἀρξάντων. An interrogative depending on an adv., as here on *ἀσαφῶς*, is unusual but quite intelligible. After *ἀρξάντων* another *πολεμοῦνται* must be supplied. "They are involved in war without any certainty which side having begun it (they are involved in war)," i.e., "which side began it."

3. **γνῶτε,** understand *κατάλυσιν*, "decide for ending the war."

Λακεδαιμονίοις...βιασαμένοις. This sentence contains a perfect labyrinth of cases. Grammatically either *Λακεδαιμονίοις* or *ὑμῖν* might be the dat. after *ἔξεστιν*, "it is possible for," but the sense shows that it must be *ὑμῖν*. *φίλους* refers to *ὑμῖν*; the change from the dat. to the accus. is quite common in sentences depending on impers. verbs. But in *χαρισαμένοις* we return to the dat. again. *αὐτῶν προκαλεσαμένων* refers to *Λακεδαιμονίοις* and should properly be in the dat., but the gen. absol. is used instead. Translate:—"You may (it is possible for you to) become friends of the L. on a sure basis (*βεβαίως*) at their own invitation (they themselves having invited you), doing them a favour rather than putting constraint upon them."

4. **ταὐτὰ λεγόντων,** "if we say the same things," i.e., "agree."

τὰ μέγιστα, neut. pl. as adv., "highly."

CHAPTER XXI

1. **διδομένης δὲ εἰρήνης**, "but now that peace was offered." The accus. as object of *δέξεσθαι* would be more correct grammatically than the gen. absolute.

2. **ἑτοίμους**, agreeing with *σπονδάς*. The fem. of *ἑτοῖμος* is the same as the masc. in Thucydides.

3. **Κλέων.** Cleon has been mentioned before by Thucydides (III. 36) as a violent but fortunately unsuccessful advocate of the proposal to punish the Mytilenaeans, who had revolted from Athens, by putting all the men to death and enslaving the women and children. He is referred to in that passage as *βιαιότατος τῶν πολιτῶν* ("most violent of the citizens") *τῷ τε δήμῳ παρὰ πολὺ ἐν τῷ τότε πιθανώτατος*. After the capture of Sphacteria he continued to be the leader of the war party till 422, when he went to oppose Brasidas in Thrace and was defeated and killed in a battle near Amphipolis. His reputation has suffered considerably from the fact that he made enemies of Thucydides—whose banishment is said to have been due to Cleon—and Aristophanes, who was prosecuted by him and retaliated by satirizing him in the comedy of the *Knights*. This play, produced in 424, just after Cleon's victory at Sphacteria, represents him as the worst kind of demagogue, flattering and robbing the people at the same time. It seems to have been his energy as much as his brutality that made him offensive to the conservative party represented by Nicias, to which Thucydides and Aristophanes belonged. But in spite of his obvious antipathy to the man, which has influenced most readers, Thucydides was too impartial to misrepresent seriously the character even of an enemy, and it is impossible to whitewash the author of the first decree against Mytilene.

ὡς χρὴ...κομίσασθαι, "that the men on the island must first surrender their arms and themselves and be brought to Athens, and then when they came the L. must give up... (When they had done this) they might recover the men...." Lit. "the Spartans having given up...might recover...."

Νίσαιαν καὶ Πηγὰς καὶ Τροιζῆνα καὶ Ἀχαΐαν. These former possessions of Athens in the Peloponnese were surrendered by her at the Thirty Years Peace in 445. Nisaea, on the Saronic gulf, which lies east of the isthmus of Corinth, was the harbour of Megara; Pegae, on the Corinthian gulf, was another harbour in the Megarid, about 14 miles from Nisaea on the other side. These two towns had been given up to Athens by the Megarians in 455. Troezen, a town nearly two miles from the coast in the south-eastern coast of Argolis, had been captured by the Athenians in 456. Achaea, the northernmost province of the Peloponnese, along the Corinthian gulf, had joined Athens in 453, but as it was an independent state it is difficult to see what the restoration of it to Athens would mean, except perhaps the renewal of the alliance.

Ἀθηναίων...σπονδῶν, "when the Athenians had made these concessions owing to their disasters and because they needed peace then somewhat more (than the Spartans)."

CHAPTER XXII

1. **πείθωσιν ἀλλήλους**, "persuade one another to accept."

2. **πολὺς ἐνέκειτο**, "attacked them (came down upon them) violently." This adverbial use of the adj. *πολύς* is common, and is applied especially to torrents and gales, as with *ῥέω*, "flow." The phrase *πολλὸς ἐνέκειτο λέγων* is used by Herodotus, VII. 158, to describe a violent speech.

οἵτινες, "seeing that they."

τῷ μὲν πλήθει. Cleon's objection to secret diplomacy was probably due to his fear that the peace party represented by the pro-Spartan Nicias might be the stronger on a small committee, whereas in open debate his own great influence over the people would prevent them from making peace.

ὀλίγοις. The dat. is explained by the *ξυν* in *ξύνεδροι*.

3. **εἴ τι καὶ...ξυγχωρεῖν**, "even if they did think fit to make any concession."

καὶ οὐ τυχόντες, "and not having succeeded," i.e., "without success."

CHAPTER XXIII

1. **διελέλυντο.** The plpf. implies that the ending of the treaty was simultaneous with the return of the ambassadors; the aor. tense would have meant that the one event followed the other.

καθάπερ ξυνέκειτο, "as was in the agreement."

ὅτι δή..., "that, to use their own words..." *δή* is slightly ironical.

ἀδίκημα ἐπικαλέσαντες τὸ τῶν νεῶν, "accusing them of injustice in refusing to restore the ships," lit. "bringing up (*ἐπικαλέσαντες*) the matter of the ships against them as an injustice."

2. **τὰ περὶ Πύλον...ἐπολεμεῖτο**, "the war at Pylus was carried on."

ἐναντίαιν, on opposite sides of the island.

σκοποῦντες...παραπέσοι, "looking for any opportunity that might occur." *εἰ* with an indefinite pron. or adv. followed by the opt. often means "in the expectation that" or "in the hope that" something may happen. Cf. *εἴ πως...ἕλοιεν* in ch. XI.

CHAPTER XXIV

1. **τὸ ἄλλο ναυτικόν**, referred to in ch. I.

4. **ἤλπιζον...χειρώσεσθαι...γίγνεσθαι.** Note the change of tense; "believed that they *would* take...and that their position *was becoming* strong."

οὐκ ἂν εἶναι, impers., "it would not be possible."

5. **βραχύτατον.** The distance between the two towns is about six miles, but this is not the narrowest space between the two sides of the straits.

ἡ Χάρυβδις. Scylla and Charybdis were legendary monsters of the sea, once women, who lived on two rocks in the straits of Messina. Charybdis caused a huge whirlpool by sucking down the waters of the sea and spouting them out again three times a day; Scylla with her six hands caught six sailors from every passing ship and devoured them in her cave. How Odysseus passed them is told in Homer's *Odyssey* (XII. 234–259), where Charybdis is thus described:—"On the other side divine Charybdis sucked down in horrible wise the salt waters of the sea. When she poured them forth, like a cauldron over a great fire she surged up all roaring in a tumult of waves, and the foam rising on high fell upon the summits of the two rocks. But when she sucked down again the salt waters of the sea she was seen all full of turbulent waves within; the rock roared horribly around her, and the earth appeared beneath her dark with sand." To pass near Charybdis was fatal to a ship and the whole crew; Odysseus, who had been warned before by Circe, kept nearer to Scylla, on the Italian side, and thus lost only six of his men.

διὰ...στενότητα καὶ...ἐσπίπτουσα..., "on account of its narrowness and because it enters...."

Τυρσηνικοῦ...Σικελικοῦ, names given to the parts of the Mediterranean along the western coast of Italy and the eastern coast of Sicily respectively.

CHAPTER XXV

1. **τῷ μεταξύ**, "the space between," i.e. "the straits."

ἠναγκάσθησαν. The word seems a strange one here, as the Sicilians *ναυμαχίας ἐβούλοντο ἀποπειρᾶσθαι*, and were superior in numbers. It must mean only that they had not chosen this particular time and place for a fight.

περὶ πλοίου διαπλέοντος, "for a ship that was sailing through," evidently a merchant ship belonging to the Syracusans or their allies.

2. **ὡς ἕκαστοι ἔτυχον**, "as each detachment happened to be," "each detachment as best it could."

ἐς τὰ οἰκεῖα στρατόπεδα. The Syracusans and Locrians had become separated, the former sailing to their station at Messana, and the latter to theirs in the Rhegine territory (*ἐν τῷ Ῥηγίῳ*), which they had invaded (ch. xxiv.).

3. **Πελωρίδα**, Peloris, now the Capo di Faro, a promontory forming the north-eastern corner of Sicily, 8 miles from Messana. *τῆς Μεσσήνης*, "in the territory of Messana."

4. **χειρὶ σιδηρᾷ ἐπιβληθείσῃ**, "by a grappling-iron (lit. "an iron hand") that was cast upon it." As the Syracusan ships were empty this must have been done by soldiers (*ὁ πεζός*) from the land. That this was not impossible is proved by the part taken by the Spartan soldiers in the sea-fight at Pylus (ch. xiv.).

5. **ἀποσιμωσάντων**, "having made a blunt edge," or obtuse angle, with their former course, i.e., "having swerved aside," away from the coast, along which they had been sailing, and out to the open sea.

6. **οὐκ ἔλασσον ἔχοντες**, lit. "having none the worst of it," "having been by no means worsted." But *οὐ* with a comp. of inferiority generally indicates superiority in Greek; "having had the best of it."

7. **Καμαρίνης**. Camarina was a city on the south coast of Sicily, forty miles from Cape Pachynum. Though it was a Dorian

city, like Syracuse, it took the side of the Leontines in their war with the Syracusans, owing to jealousy of the latter. But there was a party in Camarina which desired to bring their city into the Syracusan alliance.

προδίδοσθαι, "was being offered by treachery," "the surrender of Camarina was being contrived." Cf. for the present tense *προδιδομένην* in ch. VII.

ἐν τούτῳ, after the departure of the Athenian fleet.

Νάξον τὴν Χαλκιδικήν. Naxus, a town on the east coast of Sicily, south of Messana, was founded in 735 by settlers from Chalcis in Euboea. It was at this time an ally of Athens.

8. **τειχήρεις ποιήσαντες**, "having driven (lit. "made") them within their walls."

Ἀκεσίνην. The river Acesines, now Alcantara, at the mouth of which Naxus lies.

9. **Σικελοί**, those inhabitants of Sicily who originally came from Italy, before any of the Greek settlers, called Siceliots, came to the island. As non-Greeks the Sicels are called below *βάρβαροι*.

ὑπὲρ τῶν ἄκρων...κατέβαινον, "came down over the hills"; but *ὑπέρ* with the gen. is unusual with a verb of motion, and *οἱ ὑπὲρ τῶν ἄκρων*, "the Sicels beyond the hills," would be an easier reading.

Λεοντῖνοι, enemies of the Syracusans and allies of Athens. Their city Leontini lay between Syracuse and Naxus, eight miles from the sea.

10. **αἱ νῆες...διεκρίθησαν.** Only Messanian ships have been mentioned, but *ἐπ' οἴκου ἕκασται* shows that allied ships must have accompanied them.

ὡς κεκακωμένην. *ὡς* with the ptcp. refers, as usual, to the thoughts or feelings of the subject of the sentence; "confident that it had been (as having been) seriously weakened."

οἱ μὲν Ἀθηναῖοι. The Athenian force was purely naval, and is therefore contrasted with *ὁ πεζός*.

ἐπείρων, "assailed it."

ὁ δὲ πεζός. *προσβάλλων* must be understood from *προσβάλλοντες* above.

12. **ἐστράτευον**, "continued to make war." The Sicilians had feuds of their own, apart from their friendship or enmity towards Athens. But in the spring of the next year, 424, a congress of the

Greek states in Sicily was held at Gela, and on the proposal of the Syracusan Hermocrates a general peace was made. The Athenians were included in this, and their fleet returned home, much to the indignation of the people (IV. 58–65).

CHAPTER XXVI

2. **ὅτι μή**, "except."

οἷον εἰκὸς ὕδωρ. Supply *πίνειν*, depending on *εἰκός*, "the sort of water it was natural to drink" from the shingle, "the sort of water that might be expected."

3. **σῖτον ἐν τῇ γῇ ᾑροῦντο.** The Greeks generally landed from their ships to take their meals, a fact which accounts for their tendency to keep close to the shore. See ch. xxx.

4. **ὁ χρόνος...ἐπιγιγνόμενος**, "the prolonging (coming on) of the time beyond their expectation." *παρὰ λόγον = παρὰ λογισμόν*, "contrary to calculation." The ptcp. must sometimes be translated as a noun itself, and the noun in the nom. with which it agrees must be translated as though it were in the gen. case. So *οἱ Λακεδαιμόνιοι προειπόντες* below.

οὓς ᾤοντο. *οὕς* has no antecedent, but in the previous clause the words "(the time) involved in the capture of those men" must be understood. Translate:—"since they expected that they would capture them" (*οὕς*, the Spartans).

ἡμερῶν ὀλίγων, "within a few days."

5. **οἱ Λακεδαιμόνιοι προειπόντες**, "the fact that the L. proclaimed," or "the proclamation of the L.," the same use of the ptcp. and nom. as above. This explains the neut. sing. *αἴτιον*; it is as though Thucydides had written *τὸ τοὺς Λακεδαιμονίους προειπεῖν*.

τὸν βουλόμενον, "he who wished," "anyone who wished."

εἴ τι ἄλλο βρῶμα (τοιούτων βρωμάτων) οἷα...ξυμφέρῃ, "any other kind of food (that they chose to bring in) of all such as are useful for a siege," i.e., "to sustain a siege."

ἀργυρίου πολλοῦ, gen. of price, like *χρημάτων* below, "fixing the price at a large sum."

6. **ἄλλοι τε...καὶ μάλιστα.** The Greeks say "others and the Helots" where we should say "the Helots and others." Translate:—"The Helots and others brought in food..., the Helots especially."

ὁπόθεν τύχοιεν, "from wherever they happened (to start)," i.e., "from any place they chose." *τύχοιεν* is the opt. of indefinite frequency with relat. prons. and advs. So *ὁπότε πνεῦμα εἴη* and *ὅσοι κινδυνεύσειαν* below.

7. **ἐτήρουν**, "they waited (for an opportunity) to...."

τοῖς δὲ ἀφειδὴς...καθειστήκει, "the sailing had become unaffected by regard to the cost." The meaning of the adj. applies rather to *τοῖς δέ*, "had been made such that *they* were indifferent to losses," "had been arranged on such terms that they were..."

9. **ἐτεχνῶντο**, followed first by a prolate inf. and then by an object clause, i.e., an accus. and inf. construction ; but the accus. has to be supplied, either from *οἱ μέν* or *τὰ σ.τια*. "They devised means, the one side to send in food and the other that they (or "this") should not escape their observation."

CHAPTER XXVII

1. **σῖτος...ὅτι ἐσπλεῖ.** The peculiar order of the words, with the conj. *ὅτι* coming after *σῖτος*, the subject of its clause, is caused by the desire to balance this clause with that which precedes, the subject in each case coming first and the *ὅτι* just in front of the verb.

χειμών, "winter," not "rough weather," as is shown by *οὐδ' ἐν θέρει* below.

ἐν χωρίῳ ἐρήμῳ. These words are left very much in the air, and should have some ptcp. to depend on, e.g. *ὄντων ἐκείνων*, "as they (the men at Pylus) were in a desolate region." But Thucydides occasionally omits the pres. ptcp. of *εἰμί*. Cf. *ὡς ἐπ' ἀξιόχρεων (ὄν)* in ch. xxx. and *ἐν ταῖς εὐναῖς ἔτι (ὄντας)* in ch. xxxii.

οὐκ ἐσόμενον, "would not be possible."

περιγενήσεσθαι. Thucydides begins with the accus. and ptcp. (*κομιδὴν...ἐσομένην*) after verbs of perceiving, for the impossibility of sending food and of anchoring in winter was a *fact*; but the escape of the Spartans was only a possibility believed in by the Athenians, and the sentence therefore passes to the accus. and inf. construction after verbs of thinking, as though *ὁρῶντες* were *νομίζοντες*.

2. **ἐφοβοῦντο μάλιστα...**, for *μάλ. ἐφ. ὅτι ἐνόμιζον τοὺς Λακεδαιμονίους...* Do not translate, "they feared the Spartans most," but, "they feared most because they thought that the Spartans..."

ἔχοντάς τι ἰσχυρόν contains the main point of the clause:

"that it was owing to their possession of some strong ground for confidence," lit. "that it was as possessing something strong."

3. **ᾑρέθη**, aor. pass. of *αἱροῦμαι*, "choose," mid. of *αἱρῶ*.

4. **φανήσεσθαι**, a change from the *ὅτι* clause after *γνούς* to the inf. construction used after verbs of thinking. *γιγνώσκω*, a verb of perceiving, usually takes the ptcp., but a third ptcp. following *γνούς* and *εἰπών* and depending on the former would be very awkward.

ὡρμημένους τι τὸ πλέον...στρατεύειν, "somewhat the more enthusiastic for the expedition." *τὸ πλέον* probably = "more," like *τοῦ...πλέονος* in ch. XXI., but it might mean "for the most part."

5. **Νικίαν.** Nicias was the leader of the conservative party, or moderate democrats, who were inclined to peace with Sparta and hated Cleon. The peace of 421, which followed almost immediately after the death of Cleon, was mainly the work of Nicias, and was called by his name. As a general he gained some successes, including the capture of Minoa in 428 and of Cythera in 424, but he was generally considered much too cautious, as may be seen from the verb coined by Aristophanes, *μελλονικιᾶν*, "to wait and see" (lit. "to delay victory" or "delay like Nicias"). His hesitation and also his superstition were responsible for the great disaster at Syracuse in 413, when the whole Athenian army was captured and he himself and Demosthenes were put to death by the Syracusans. Thucydides says "he least of all men was worthy to meet such a fate, owing to his practice of every traditional virtue"—a phrase undoubtedly not ironical. Nicias was a victim of the extraordinary Athenian habit of electing politicians as generals.

ῥᾴδιον εἶναι. The inf. depends on the idea of "saying" implied in *ἀπεσήμαινεν*. *ἄν* might be expected with the inf., but "was easy" comes to much the same as "would be easy."

εἰ ἄνδρες εἶεν οἱ στρατηγοί, "if the generals were *men*."

CHAPTER XXVIII

1. **τὸ ἐπὶ σφᾶς εἶναι**, "as far as they (he and the other generals) were concerned."

2. **λόγῳ**, "in pretence."

ἑτοῖμος, "ready to accept."

παραδωσείοντα (formed from *παραδίδωμι*), one of a few verbs with the termination *-σείω*, which implies a wish or willingness.

οὐκ ἔφη αὐτὸς...στρατηγεῖν, "said that not he but Nicias was commander." This is a good instance of the construction after verbs of saying or thinking. When the subject of the inf. is the speaker himself any adj., ptcp., or pron. referring to him is in the nom. case, as *αὐτός*; when it is anyone else the accus. is used, as *ἐκεῖνον*. *οὔ φημι*, not *φημὶ οὐ*, for "say that...not."

οὐκ ἂν οἰόμενος...ὑποχωρῆσαι. The *ἂν* goes with *τολμῆσαι*, "not thinking that he would have consented to resign in his favour" (*οἱ*). *τολμῶ* has a wider sense than "dare"; it can mean "have the heart to" or, as here, "be ready to." *οἱ*, "to him," dat. of reflex pron. *ἕ*.

3. **ἐξίστατο**, impf., "offered to resign," lit. "stand out of."

τὰ εἰρημένα, "what he had said," a strange accus. after an intr. verb with the prep. *ἐξ*.

4. **ἐξαπαλλαγῇ**, deliberative subjunct. after *οὐκ ἔχων*, "not knowing (lit. "having") how he was to escape from what he had said."

Λημνίους καὶ Ἰμβρίους. Lemnos and Imbros, two islands in the north of the Aegean, belonged to Athens during the greater part of the fifth and fourth centuries B.C.

Αἴνου, still called Enos, a town on the coast of Thrace, near the Dardanelles. Thrace on several occasions supplied Athens with light-armed troops (*πελτασταί*).

5. **ἀσμένοις...ἐγίγνετο τοῖς σώφροσι**, "sober-minded men were glad of it," lit. "it happened to sober-minded men (so that they were) glad." This is a common construction in Greek; cf. *βουλομένῳ μοί ἐστιν*, "I am willing."

σφαλεῖσι is conditional; "or that he (Cleon) would capture the Spartans for them, *if* they were deceived in (of) their expectation."

CHAPTER XXIX

2. **τὸν δὲ Δημοσθένη.** According to Aristophanes (*Knights*, 54–57) all the credit for the success at Sphacteria was due to Demosthenes.

ὥρμηντο, plpf., "had been made eager," "were eager."

ῥώμην, "strength," here "strength of *moral*, confidence."

ἡ νῆσος ἐμπρησθεῖσα, "the island burnt," = "the burning of the island," i.e., "a fire on the island." For the construction see note on ch. XXVI., *ὁ χρόνος...ἐπιγιγνόμενος*.

3. **ἐφοβεῖτο**, "he was afraid," intransitive.

ἐνόμιζε. This verb, repeated below, governs all the accus. and inf. constructions in the rest of this chapter.

στρατοπέδῳ, dat. governed by *προσβάλλοντας*. *ἀποβάντι* is conditional ("if a large army landed"), and *ἄν* must be taken with *βλάπτειν*, "the enemy, attacking it, would inflict on it great losses."

δῆλα, neut. pl. after two fem. nouns.

τοῦ δὲ αὐτῶν, "of their army," i.e., the Athenians.

ἐπ' ἐκείνοις...ἐπιχείρησιν, "for the initiative (attack) would be in their power," "lie with the Spartans."

4. **εἰ δ' αὖ...βιάζοιτο**, "again, if he were forced..."

τοὺς ἐλάσσους ἐμπείρους δέ, "the force that was smaller, but acquainted with the ground."

διαφθειρόμενον goes with *λανθάνειν*, "would be destroyed without his observing it, since it was so large."

τῆς προσόψεως, "the possibility of seeing."

CHAPTER XXX

1. **τοῦ Αἰτωλικοῦ πάθους**, the unsuccessful invasion of Aetolia by Demosthenes (III. 97, 98). Thucydides mentions that many of the defeated Athenians lost their way and entered a wood, which was set on fire by the Aetolians. It is strange that the enterprising Demosthenes did not do the same to the wood on Sphacteria, without waiting for it to happen by accident.

μέρος τι, "to a certain extent," the limiting accusative.

2. **διὰ προφυλακῆς**, "with an advanced outpost."

κατὰ μικρόν = *μικρόν τι*, forming the object of *ἐμπρήσαντος*. Cf. *ἐπὶ πολὺ τῆς χώρας* in ch. III. and *ἐπὶ πολὺ ἐποίει* in ch. XII.

ἀπὸ τούτου, "*after* this."

3. **ὑπονοῶν πρότερον**, explaining *μᾶλλον κατιδών*, "for he suspected before...."

αὐτοῦ. If the reading of the MSS. is correct the sentence must be translated "for fewer men than it" (the food); i.e., "for fewer men than the number of rations which were supplied." But *αὐτούς* or *αὐτοῖς* is a much more natural reading.

ὡς ἐπ' ἀξιόχρεων (neut.), "as for an object worthy that the Athenians should display more energy," "that deserved a greater display of energy from the Athenians." τοὺς Ἀ. σπουδὴν ποιεῖσθαι is an accus. and inf. construction depending on ἀξιόχρεων, like the inf. after ἄξιόν ἐστι, "it is worth while"; but it would be easier to read τοῖς Ἀθηναίοις.

4. **ὡς ἥξων.** ὡς with the fut. ptcp. after a verb of action = "in order to *carry out* an intention"; after a verb of information (προπέμψας ἄγγελον = ἀγγείλας) the words mean "in order to *announce* an intention." Translate, "to say that he was coming."

ἅμα γενόμενοι, "having met."

φυλακῇ τῇ μετρίᾳ, "custody that would not be too strict." The article τῇ implies "*the* form of custody that would be reasonable."

περὶ τοῦ πλέονος, "about the larger matter," i.e., a general peace.

CHAPTER XXXI

1. **τοὺς ὁπλίτας πάντας.** These would be the 10 or 12 ἐπίβαται (marines) on each of the 70 ships, 800 in all.

ὀλίγον, "a little," goes with πρὸ τῆς ἕω.

ἐκ...τοῦ πελάγους, "on the side of the open sea." So ἐκ θαλάσσης καὶ ἐκ τῆς γῆς below.

πρὸς τοῦ λιμένος, "facing the harbour."

διετετάχατο. The usual form of the plpf. would be διατεταγμένοι ἦσαν.

2. **μέσον,** i.e., τῆς νήσου.

εἰ καταλαμβάνοι, "if a precipitate retreat were forced upon them (came upon them)." βιαιοτέρα, "followed up with great violence"; the comp. merely intensifies the adj., "more violent than in an ordinary attack."

CHAPTER XXXII

1. **λαθόντες τὴν ἀπόβασιν,** "having carried out the landing unobserved," a strange use of the accus. after λανθάνω; usually the object is the person by whom the action is not observed.

2. **ἀπέβαινον,** pl. because στρατός is a collective noun; but the sing. is more regular with such nouns.

θαλαμιῶν. The θαλαμιοί or θαλαμῖται were the 54 rowers on the lowest bench in a Greek trireme; there were 62 θρανῖται on the highest and 54 *ζυγῖται* on the middle bench. Thus the rowers on a Greek trireme were 170 in number, and 116 men were now taken from each of the 70 ships; with the 800 hoplites already landed and the 1600 archers and targeteers the whole number would be 10,500, not counting the Messenians and others. There were only 420 Spartans on the island.

κατεῖχον, "occupied (posts)," "were stationed."

3. **καὶ μὴ ἔχωσι...ἀντιτάξωνται**, "and they might not know whom to face," lit. "draw up their line against." *ἀντιτάξωνται*, deliberative subjunctive.

ἑκατέρωθεν, i.e., in front and in rear.

4. **ᾗ χωρήσειαν**, conditional opt., "wherever they might go."

ἀπορώτατοι, "hardest to deal with."

ἐκ πολλοῦ ἔχοντες ἀλκήν, "relying on (having their strength in) ...at a distance."

μηδέ. *μή* is the regular negative in relat. sentences referring to a class of persons.

ἀναχωροῦσιν (*τοῖς πολεμίοις*).

5. **ἔταξεν** (*τοὺς στρατιώτας*).

CHAPTER XXXIII

1. **οἱ δὲ περὶ τὸν Ἐπιτάδαν...νήσῳ**, "those with Epitadas, the main body of the Spartans on the island."

2. **ἐκεῖνοι**, the Athenian hoplites.

ᾗ...προσκέοιντο, opt. of indefinite frequency, "wherever they ran forward and attacked them." Cf. *ᾗ προσπίπτοιεν* in ch. xxxiv.

καὶ οἵ, "and they." *οἵ* is here a demonstr. pron., not relative.

προλαμβάνοντες...τῆς φυγῆς, "getting the advantage in flight." *τῆς φυγῆς* is an extension of the gen. of space traversed.

χωρίων...τραχέων ὄντων. After giving the first reason in the dat., *χαλεπότητι*, Thucydides gives the second in the gen. absol. *τραχέων ὄντων* (*χωρίων*, which goes with both constructions). *χαλεπότητι* refers to the natural ruggedness of the ground, *τραχέων* to its uncultivated state.

ὅπλα, "heavy armour," as contrasted with the light armour of the Athenian *ψιλοί* or *πελτασταί*.

CHAPTER XXXIV

1. **τοῦ θαρσεῖν...εἰληφότες**, "having gained the greatest (degree of) courage." *πολλαπλάσιοι φαινόμενοι* amplifies *τῇ ὄψει*, and gives the reason for their courage, "seeing with their own eyes that they were many times the number (of the enemy)."

ξυνειθισμένοι...φαίνεσθαι, a blending of two constructions, (1) *ξυνειθισμένοι ὥστε...φαίνεσθαι*, "becoming accustomed to them, so that they no longer appeared formidable," and (2) *ξυνειθισμένοι...(νομίζειν)*, "becoming accustomed to think them no longer formidable."

ἄξια τῆς προσδοκίας, "things worthy of their expectation," i.e., "anything that came up to their expectation."

τῇ γνώμῃ δεδουλωμένοι ὡς ἐπὶ Λακεδαιμονίους, "awed" or "cowed in spirit by the thought that (*ὡς*) they were landing to fight Spartans," whose army had as great a reputation as the Athenian navy.

2. **ἐχώρει πολὺς ἄνω**, "rose up in clouds."

3. **χαλεπὸν...καθίστατο**, "became difficult for," "went hard with."

πῖλοι. There is some doubt whether this word means "felt caps," or "felt cuirasses." As Thucydides has laid such stress on the heavy armour of the Spartans the latter rendering seems less natural. In Aristoph. Lysistr., l. 562, the adj. *χαλκοῦς* is used of the *πῖλος*, which shows that it had come to mean simply "helmet."

ἐναπεκέκλαστο, "had broken off short (in their armour) as they were struck."

βαλλομένων (*αὐτῶν*), gen. absolute.

ἀποκεκλῃμένοι...προορᾶν, "shut off in their (power of) sight from seeing ahead," i.e., "prevented from seeing ahead."

CHAPTER XXXV

2. **ἐγκατελαμβάνοντο**, "were caught in it," i.e., "in retreat."

3. **περίοδον...οὐκ εἶχον**, "had no way of going round and encircling them owing to the strength of their position."

4. **τῆς κυκλώσεως**, "*the* means of surrounding them" that existed before.

CHAPTER XXXVI

1. **περιμέναι.** The inf. is used in a final sense after verbs of giving (here *δοῦναι*) and choosing.

δοκεῖν βιάσεσθαι, "(he said) he thought he would force."

2. **τὸ ἀεὶ παρεῖκον τοῦ κρημνώδους,** "any parts of the cliff that from time to time yielded them footing." *ἀεί* is often used of a series; *ὁ ἀεὶ παρελθών*="the men who from time to time came forward," not "the man who always came forward."

3. **καὶ οἱ Λακεδαιμόνιοι...ἀντεῖχον.** The main verb is *ἀντεῖχον*, but the construction is irregular, with *γάρ* and *οὗτοί τε* coming in between.

ὡς μικρὸν μεγάλῳ εἰκάσαι, "to compare small with great."

ἐν Θερμοπύλαις. In 480 a few thousand Greeks met the huge invading army of Persians under Xerxes at Thermopylae, the famous pass leading from Thessaly into southern Greece. They held out for two days, but the Persians then discovered from a Greek traitor the existence of a mountain path (*ἀτραπός*) and made their way by it to the rear of their enemies' position. Upon this most of the Greeks retreated, at the command of the Spartan king Leonidas, but Leonidas himself with 300 Spartans and 700 Thespians remained and fought to the last. It was owing to this act of heroism that the Spartans originally gained their reputation for bravery, and that the other Greeks expected them "never to surrender their arms but fight with them to the death" (ch. XL.).

CHAPTER XXXVII

1. **ὁποσονοῦν,** neut. as adv., "ever so little."

εἴ πως...ἐπικλασθεῖεν, "(kept back their men) if perchance the Spartans might...," "in the hope that the Spartans might be broken in spirit."

2. **παραδοῦναι.** The words *ἐπικλασθεῖεν τῇ γνώμῃ* are taken as equivalent to a verb of persuading in the pass., and therefore followed by a simple infinitive. But it has been proposed to omit the words *τὰ ὅπλα παραδοῦναι*, since they also occur in the following clause, and the repetition is awkward.

ὥστε βουλεῦσαι. *ὥστε* here=*ἐφ' ᾧ*, "on condition that they (the A.) should...."

CHAPTER XXXVIII

1. **δηλοῦντες προσίεσθαι**, "signifying that they accepted."

3. **ἀφιέντων** goes with *τῶν Ἀθηναίων*, not *ἐκείνων*, which is the partitive gen. after *οὐδένα*; "none of them," the Spartans.

μηδὲν αἰσχρὸν ποιοῦντας, "provided you do nothing dishonourable."

4. **τοὺς νεκροὺς διεκομίσαντο.** Burial of the dead was considered so important by the Greeks that the defeated side always asked for a truce after a battle, in order to recover and bury the bodies of their slain soldiers. This was a confession of defeat, as the setting up of a trophy was an assertion of victory.

5. **καὶ ζῶντες**, "*or* living"; *τοσοίδε*, "the following numbers."

ὀκτὼ ἀποδέοντες τριακόσιοι, "three hundred all but (lacking) eight."

CHAPTER XXXIX

1. **ἐγένετο**, "came to." *γίγνομαι* is frequently used of numbers in this sense. Cf. ch. IX., *ὡς τεσσαράκοντα ἐγένοντο.*

2. **τοῖς ἐσπλέουσι**, understand *σιτίοις*, "on the food that was brought in by boats."

ἢ πρὸς τὴν ἐξουσίαν, "than (he might have done) in view of his resources."

3. **μανιώδης οὖσα.** Thucydides regards the promise made by Cleon as "mad" because Cleon at Athens could not know the difficulties at Sphacteria, described in ch. XXIX., and because he fixed so short a time for its performance. Thucydides, like most Greeks, was too much impressed by "the uncertainty of the future" (*τὸ ἀστάθμητον τοῦ μέλλοντος*) to approve of any such definite promise in military matters. He obviously considers Cleon's part in the success as due to luck, and luck is no proof of wisdom. For the tactics employed the whole credit is given to Demosthenes.

CHAPTER XL

1. **ἔχοντας**, i.e., *τὰ ὅπλα.*

2. **ἀπιστοῦντές τε μὴ εἶναι...ἀπεκρίνατο.** (1) This sentence contains an anacoluthon, i.e., the nom. at the beginning is left hanging in the air, without a main verb; for *ἀπεκρίνατο* has a quite different subject. The *nominativus pendens* is not uncommon in Thucydides.

(2) After verbs of doubting and denying *μή* is used with the accus. and inf., but as it only repeats the negative idea already implied in the verb it must be left out in translating. Cf. the French use of "ne" in the subord. clause after verbs of fearing, or after those of doubting and denying when preceded by a negative.

δι' ἀχθηδόνα, "to irritate him." *διά* here means "for the purpose of," a rare use of this preposition.

εἰ οἱ τεθνεῶτες...κἀγαθοί, implying that those who surrendered were not *καλοὶ κἀγαθοί*. This term originally described a class, the aristocrats, but from the Spartan's answer it is obvious that it here means "gallant men."

ἄτρακτον. The usual meaning of *ἄτρακτος* is "a spindle," but it is used in Sophocles, Phil. 290 and Trach. 714, for "arrow," as here. Possibly the word was common among the Spartans in this sense, as an expression of their contempt for the arrow as "a woman's weapon." So Brasidas, in ch. XI., describes ships as *ξύλα*.

λέγων, "meaning."

ὁ ἐντυγχάνων, "any man who chanced upon them," i.e., "chanced to be in their way."

CHAPTER XLI

1. **μέχρι οὗ τι ξυμβῶσιν**, "until they should come to some agreement," i.e., "make peace." *ἄν* is omitted after *μέχρι οὗ* as in ch. XVI., *μέχρι οὗ ἐπανέλθωσιν*.

2. **οἱ ἐκ τῆς Ναυπάκτου Μεσσήνιοι.** Naupactus had been given over by the Athenians in 459 to settlers from Messenia, who had been driven out from their own land by the Spartans.

3. **ἀμαθεῖς ὄντες...φοβούμενοι.** Three reasons are given for the dismay of the Spartans, expressed in three participles, the first and third agreeing with *οἱ Λ.* and the second belonging to a gen. absol. construction (*τῶν τε Εἱλώτων αὐτομολούντων*).

μὴ καὶ ἐπὶ μακρότερον, "lest some of the conditions in their country (*τῶν κατὰ τὴν χώραν*) should be disturbed by a revolution on a wider scale," i.e., "lest there should be a revolution on a wider scale in their country." To prevent this the Spartans next year proclaimed that all the Helots who thought they had distinguished themselves in war should come forward and claim their freedom. Out of those who did so 2,000 were chosen, apparently to be set free, and "put garlands on their heads and went round to all the temples to celebrate their liberation; but not long afterwards the Spartans made away with them, and no man knew how each was done to death" (IV. 80). Among these may have been some of the Helots who took food across to the Spartans in Sphacteria.

οὐ ῥᾳδίως ἔφερον, "did not take it calmly," "were filled with dismay."

ἔνδηλοι εἶναι, "to be manifest," i.e., "to betray their distress."

4. **φοιτώντων**. The accus. would be the natural case here; the gen. absol. is not used as a rule when the ptcp. refers to the subject or object of the main verb, as *φοιτώντων* here refers to *αὐτούς*, the object of *ἀπέπεμψαν*. With *φοιτώντων* the gen. *αὐτῶν* must be supplied.

VOCABULARY

The following parts of irregular verbs are given here to avoid a number of cross-references, as they occur in many compounds and their pres. indicatives are difficult to trace.

βάς (gen. *βάντος*), *βῆναι*, *βαίην*, aor. ptcp., inf. and opt. of **βαίνω**.

εἵς (gen. *ἕντος*), *εἶναι*, *εἵην*, aor. ptcp., ... of **ἵημι**.

ἐλθών, *ἐλθεῖν*, *ἔλθω*, *ἔλθοιμι*, aor. ptcp., inf., subjunct. and opt. of **ἔρχομαι**.

ἑλών, ..., aor. ptcp., ... of **αἱρῶ[έω]**.

ἐνεγκών, ..., aor. ptcp., ... of **φέρω**.

ἰδών, ..., aor. ptcp., ... of **ὁρῶ[άω]**.

πεσών, ..., aor. ptcp., ... of **πίπτω**.

στάς (gen. *στάντος*), *στῆναι*, *σταίην*, 2nd aor. ptcp., inf., and opt. of **ἵστημι**.

σχών, *σχεῖν*, *σχῶ*, *σχοίμι*, aor. ptcp., inf., subjunct. and opt. of **ἔχω**.

For aorists beginning *ἠ*— see under *αἰ*—.
,, ,, ,, *ᾠ*— ,, ,, *οἰ*—.

When not otherwise stated the verbs are given in the following order: pres., fut., aor., and pf. indicative.

ἀγαθ-ός, *-ή*, *-όν*, *good*, *brave*; comp. **ἀμείνων**, superl. **ἄριστ-ος**.

ἄγαν, adv. *too much*.

ἀγγεῖ-ον, *-ου*, n. *vessel* (of metal etc.).

ἀγγέλλω, *ἀγγελῶ*, *ἤγγειλα*, *announce*, *bid*.

Αγ-ις, *-ιδος*, m. *Agis* (king of Sparta).

ἄγω, *ἄξω*, *ἤγαγον*, *lead*, *bring*, *hold*.

ἀδίκημα, *-τος*, n. *wrong*, *injustice*.

ἀδόκητ-ος, *-ον*, *unexpected*; **τὸ ἀδόκητον**, *the unexpected* (*sight*).

ἀδύνατ-ος, *-ον*, *impossible*, *unable*.

ἀεί, adv. *always*, *from time to time*.

ἀήθ-ης, *-ες*, *unused to* (with gen.); adv. **ἀήθως**, *unexpectedly*.

Ἀθῆν-αι, -ων, f. pl., *Athens*.
Ἀθηναῖ-ος, -α, -ον, *Athenian*.
ἀθρό-ος, -α, -ον, *all together*.
ἀθυμί-α, -ας, f. *despondency*.
ἀΐδι-ος, -ον, *eternal*.
Αἶν-ος, -ου, f. *Aenus*.
αἱρῶ[έω], aor. εἷλον, aor. pass. ᾑρέθην, *take*; in mid. *choose*.
αἴρω, ἀρῶ, ἦρα, *raise*; intr. *set out*.
αἰσχρ-ός, -ά, -όν, *disgraceful*.
αἰσχύν-η, -ης, f. *shame*.
αἴτι-ος, -α, -ον, *responsible*; neut. αἴτιον as noun, *cause*.
αἰτῶ[έω], act. or mid. *ask for*.
Αἰτωλικ-ός, -ή, -όν, *Aetolian*.
αἰχμάλωτ-ος, -ου, m. *prisoner*.
Ἀκαρνανί-α, -ας, f. *Acarnania*.
Ἀκεσίν-ης, -ου, m. *Acesines*.
ἀκίνδυν-ος, -ον, *without risk*.
ἀκμ-ή, -ῆς, *point, highest point, ripeness*.
ἀκόντι-ον, -ου, n. *javelin*.
ἀκούω, ἀκούσομαι, ἤκουσα, *hear*.
ἄκρ-α, -ας, f. *promontory*.
ἄκριτ-ος, -ον, *undecided*.
ἀκροβολίζομαι, *skirmish at a distance*.
ἄκρ-ον, -ου, n. *height*.
ἀκρόπολ-ις, -εως, f. *citadel*.
ἀκρωτήρι-ον, -ου, n. *extremity, promontory*.
ἄκ-ων, -ουσα, -ον, *unwilling, not intending it*.
ἀληθ-ής, -ές, *true*.
ἀλίμεν-ος, -ον, *harbourless*.
ἁλίσκομαι, ἁλώσομαι, ἑάλων, *be caught*.
ἀλκ-ή, -ῆς, f. *strength, defence*.
ἀλλά, conj. *but*.
ἀλλήλ-ους, -ας, -α, (no nom.), *one another*.
ἄλλοθεν, adv. *from elsewhere*.
ἄλλ-ος, -η, -ο, *other*; adv. ἄλλ-ως, *in vain*.
ἁλμυρ-ός, -ά, -όν, *salt, brackish*.
ἄλφιτ-ον, -ου, n. *barley-meal*.
ἀλῶ[έω], pf. mid. ἀλήλεμαι, *grind*.
ἁλώσιμ-ος, -ον, *easy to take*.
ἅμα, adv. *at the same time, together*; with ptcp. *while, on*; as prep. with dat. *at, at the same time as*.
ἀμαθ-ής, -ές, *not having had experience*.
ἁμάρτημα, -τος, n. *mistake*.
ἁμαρτί-α, -ας, f. *mistake*.
ἀμείνων, comp. of ἀγαθός, *better*.
ἀμύνω, ἀμυνῶ, ἤμυνα, with acc. and dat., *ward off from, defend*; in mid. *resist*.
ἀμφίβολ-ος, -ον, *doubtful, attacked on two sides*.
ἀμφότερ-ος, -α, -ον, *each* (*of two*); in pl. *both*.
ἀμφοτέρωθεν, adv. *from both sides*.
ἄν, part. in apodosis of conditional sentences, i.e., in the clause that in Eng. contains the word *would*. With relat. and subjunct. *-ever*, e g. **ὃς ἄν**, *whoever*.
ἀναγκάζω, aor. ἠνάγκασα, *compel*.
ἀνάγκ-η, -ης, f. *necessity, compulsion, crisis*.
ἀνάγω, *lead up*; in mid. *put to sea*.

ἀναγωγ-ή, -ῆς, f. *putting to sea.*
ἀναδέω, *bind;* in mid. *take in tow.*
ἀναιρῶ[έω], *take up.*
ἀνακόπτω, aor. pass. ἀνεκόπην, *beat back.*
ἀναλαμβάνω, *take up.*
ἀνάπαυσ-ις, -εως, f. *relief, respite.*
ἀναπαύω, *rest, draw off* (from attack).
ἀνασείω, *wave.*
ἀνασπῶ, aor. ἀνέσπασα, *draw to land.*
ἀναστρέφω, *turn round;* in mid. *be, move about.*
ἀναφαίνω, aor. pass. ἀνεφάνην, *show;* in mid. or pass. *appear.*
ἀναχώρησ-ις, -εως, f. *retreat, retirement.*
ἀναχωρῶ[έω], *retreat, draw back.*
ἄνεμ-ος, -ου, m. *wind.*
ἄνευ, prep., with gen. *without.*
ἀνήκεστ-ος, -ον, *irreparable.*
ἀνήρ, ἀνδρός, m. *man.*
ἀνθέλκω, impf. ἀνθεῖλκον, *pull on the other side.*
ἀνθησσῶμαι[άομαι], *give way in turn.*
ἄνθρωπ-ος, -ου, m. *man, human being.*
ἀνίημι, aor. ἀνῆκα, *give up, relax.*
ἀνοκωχ-ή, -ῆς, f. *armistice.*
ἀνταιτῶ[έω], *ask in return.*
ἀνταλλάσσω, pf. ptcp. mid. ἀντηλλαγμένος, *change.*
ἀνταμύνομαι, *resist, retaliate.*
ἀντανάγομαι, *put out to sea against.*
ἀνταποδίδωμι, *give, show in return.*
ἀντεῖπον, *said in reply.*
ἀντεκπλέω, *sail out against.*
ἀντεπανάγομαι, *put out to sea against.*
ἀντέπειμι, *advance to meet.*
ἀντέχω, *hold out, endure.*
ἀντί, prep., with gen. *in return for, instead of.*
ἀντιλέγω, *speak against, object.*
ἀντίπαλ-ος, -ον, *equal to, balancing.*
ἀντίπρῳρ-ος, -ον, *with the prow towards.*
ἀντιτάσσω, *draw up against;* in mid. *meet in battle, face.*
ἄνω, adv. *up, upwards.*
ἀξιόλογ-ος, -ον, *worth mention.*
ἄξι-ος, -α, -ον, *worth, worthy of.*
ἀξιόχρε-ως, -ων, *worthy, considerable.*
ἀξιῶ[όω], *require, demand, expect.*
ἀξίωμα, -τος, n. *reputation.*
ἀξύνετ-ος, -ον, *unintelligent.*
ἄοπλ-ος, -ον, *weakly-armed.*
ἀπαγγέλλω, aor. ἀπήγγειλα, *bring a message.*
ἀπαίρω, *start from.*
ἀπαιτῶ[έω], *demand back.*
ἀπαλλάσσω, aor. pass. ἀπηλλάγην, *rid of.*
ἅπας, ἅπασα, ἅπαν, *all.*
ἄπειμι, *be away.*
ἀπείργω, aor. ἀπεῖρξα, *keep back.*
ἄπειρ-ος, -ον, *unacquainted with.*
ἀπέραντ-ος, -ον, *endless.*
ἀπερισκέπτως, adv. *without reflection.*

ἀπέρχομαι, aor. *ἀπῆλθον*, *go away.*
ἀπέχω, *be distant.*
ἄπιστ-ος, *-ον*, *distrustful* (with dat. *towards*).
ἀπιστῶ[έω], *disbelieve, be doubtful.*
ἄπλοι-α, *-ας*, f. *impossibility of sailing.*
ἀπό, prep., with gen. *from.*
ἀποβάθρ-α, *-ας*, f. *landing-ladder.*
ἀποβαίνω, aor. *ἀπέβην*, *land, be fulfilled.*
ἀποβάλλω, *cast away, lose.*
ἀποδέω, *lack.*
ἀποδίδωμι, *give back.*
ἀποθνήσκω, aor. *ἀπέθανον*, *die, be killed.*
ἀποικί-α, *-ας*, f. *colony.*
ἀποκληρῶ[όω], *choose by lot.*
ἀποκλῄω, pf. mid. *ἀποκέκλῃμαι*, *shut off, prevent.*
ἀποκνῶ[έω], *hesitate, hold back.*
ἀποκολυμβῶ[άω], *dive away.*
ἀπόκρημν-ος, *-ον*, *precipitous.*
ἀποκρίνομαι, aor. *ἀπεκρινάμην*, *answer.*
ἀπόκρισ-ις, *-εως*, f. *answer.*
ἀποκτείνω, *ἀποκτενῶ*, *ἀπέκτεινα*, *kill.*
ἀπολαμβάνω, *cut off.*
ἀπολέγω, *choose out.*
ἀποπειρῶμαι[άομαι], *try, venture on* (with gen.).
ἀποπέμπω, *send away.*
ἀποπίπτω, *fall off.*
ἀποπλέω, *sail away.*
ἀπορί-α, *-ας*, f. *want, difficulty.*
ἄπορ-ος, *-ον*, *difficult, impossible, hard to deal with.*
ἀπορῶ[έω], *be at a loss.*
ἀποσημαίνω, *allude.*
ἀποσιμῶ[όω], *turn aside.*
ἀποστέλλω, aor. *ἀπέστειλα*, aor. pass. *ἀπεστάλην*, *send.*
ἀποφαίνω, *point out, declare.*
ἀποχωρῶ[έω], *go away, depart.*
ἄπρακτ-ος, *-ον*, *without success.*
ἀπροσδόκητ-ος, *-ον*, *unexpected, sudden.*
ἄρα, part. *after all.*
ἀργύρι-ον, *-ου*, n. *money, silver.*
ἀρετ-ή, *-ῆς*, f. *virtue, generosity, courage.*
ἀριστοποιοῦμαι[έομαι], *take a meal.*
ἀρκῶ[έω], *be sufficient.*
ἄρτι, adv. *lately.*
ἀρχαῖ-ος, *-α*, *-ον*, *ancient;* **τὸ ἀρχαῖον**, *of old.*
ἀρχ-ή, *-ῆς*, f. *command, beginning.*
Ἀρχί-ας, *-ου*, m. *Archias.*
Ἀρχίδαμ-ος, *-ου*, m. *Archidamus.*
ἄρχω, *rule, be in command, begin* (with gen.); **ὁ ἄρχων**, *commander, magistrate.*
ἀσαφ-ής, *-ές*, *uncertain.*
ἀσθένει-α, *-ας*, f. *weakness.*
ἀσθεν-ής, *-ές*, *weak.*
Ἀσίν-η, *-ης*, f. *Asine.*
ἀσκ-ός, *-οῦ*, m. *skin, bag.*
ἄσμεν-ος, *-η*, *-ον*, *glad.*
ἀσπ-ίς, *-ίδος*, f. *shield.*
ἄτρακτ-ος, *-ου*, m. *spindle, arrow.*
ἀτραπ-ός, *-οῦ*, f. *path.*
ἀτριβ-ής, *-ες*, *pathless, untrodden.*
Ἀττικ-ός, *-ή*, *-όν*, *Attic, Athenian;* **ἡ Ἀττική**, *Attica.*

αὖ, adv. *again.*
αὖθις, adv. *again.*
αὐλίζομαι, aor. ηὐλισάμην, *pass the night, encamp.*
αὐτόθι, adv. *there, on the spot.*
αὐτομολῶ[έω], *desert, run away.*
αὑτ-όν, -ήν, -ό, contr. from ἑαυτόν.
αὐτ-ός, -ή, -ό, pron. *-self;* (only in accus., gen. and dat.) *him, her, it, them;* **ὁ αὐτός**, *the same;* **αὐτοῖς ἄνδρασι**, *with the crew.*
αὐτόσε, adv. *to that place.*
ἀφαν-ής, -ές, *unseen, hidden;* **ἐκ τοῦ ἀφανοῦς**, *from a place out of sight.*
ἀφειδ-ής, -ές, *without sparing, regardless of the cost.*
ἀφίημι, ἀφήσω, ἀφῆκα, *give up, let go.*
ἀφικνοῦμαι[έομαι], ἀφίξομαι, ἀφικόμην, ἀφῖγμαι, *arrive.*
ἀφίστημι, *put away, draw away;* in 2nd aor. ἀπέστην and mid. *revolt from* (with gen.).
ἄφνω, adv. *suddenly.*
ἀφορῶ[άω], *look at.*
Ἀχαΐ-α, -ας, f. *Achaea.*
ἀχθηδ-ών, -όνος, f. *annoyance.*

βάλλω, βαλῶ, ἔβαλον, *throw, strike, pelt.*
βάρβαρ-ος, -ου, m. *barbarian.*
βασιλ-εύς, -έως, m. *king.*
βέβαι-ος, -ον, *trustworthy, secure.*
βί-α, -ας, f. *force.*
βιάζω, aor. ἐβίασα, act. or mid. *force, drive to retreat, overpower.*
βίαι-ος, -α, -ον, *forced, violent.*
βλάπτω, *inflict harm,* or *loss, on.*
βο-ή, -ῆς, f. *shout.*
βοήθει-α, -ας, f. *coming to the rescue, succour.*
βοηθῶ[έω], with dat. *go to the aid of;* with ἐπί, *march upon, march against.*
Βοττιαῖ-οι, -ων, m. pl. *Bottiaeans.*
βουλεύω, act. or mid., *take counsel, resolve.*
βούλομαι, βουλήσομαι, ἐβουλήθην, *wish.*
βοῶ[άω], *shout, cry.*
βραδ-ύς, -εῖα, -ύ. *slow.*
Βρασίδ-ας, -ου, m. *Brasidas.*
βραχ-ύς, -εῖα, -ύ, *short;* in pl. *few.*
βρῶμα, -τος, n. *food.*
βύζην, adv. *closely.*

γαλήν-η, -ης, f. *calm weather.*
γάρ, conj. *for.*
γε, part. *at least.*
γέλ-ως, -ωτος, m. *laughter.*
γῆ, γῆς, f. *earth, land.*
γίγνομαι, γενήσομαι, ἐγενόμην, γεγένημαι, or γέγονα, *become, be, happen, come to.*
γιγνώσκω, γνώσομαι, ἔγνων, *learn, observe;* in aor. *know.*
γνούς, aor. ptcp. of γιγνώσκω.
γνώμ-η, -ης, f. *judgment. expectation;* **παρὰ γνώμην**, *contrary to expectation.*

δαπανῶ[άω], *spend, put to expense.*
δασ-ύς, -εῖα, -ύ, *bushy, woody.*

δέ, conj. *but, and,* (preceded by *μέν*) *on the other hand.*
δέδια, *δέδοικα*, pfs. of *δείδω* with pres. sense, *fear.*
δειν-ός, *-ή*, *-όν*, *terrible, formidable, clever;* **τὸ δεινόν**, *the danger.*
δεινότ-ης, *-ητος*, f. *threatening appearance.*
δέκα, *ten.*
δέχομαι, aor. *ἐδεξάμην*, *receive.*
δέω, *δεήσω*, *ἐδέησα*, aor. pass. *ἐδεήθην*, *lack, be needed;* impers. *δεῖ*, *it is necessary;* in mid. and pass. *ask, need* (with gen.); **τὸ δέον**, *what is wanted, needed.*
δή, part. *indeed;* sometimes= *to use (their) words.*
δῆλ-ος, *-η*, *-ον*, *clear, visible.*
δηλῶ[όω], *show, signify.*
δήλωσ-ις, *-εως*, f. *meaning, explanation;* **δήλωσιν ποιοῦμαι**, *signify.*
δημαγωγ-ός, *-οῦ*, *leader of the people.*
Δημοσθέν-ης, *-ους*, m. *Demosthenes.*
Δημοτέλ-ης, *-ους*, m. *Demoteles.*
δῃῶ[όω], *lay waste.*
διά, prep., with accus. *on account of;* with gen. *by means of, with, at an interval of.*
διαβαίνω, 2nd aor. *διέβην*, *cross.*
διαβάλλω, aor. pass. *διεβλήθην*, *misrepresent, slander.*
διαβιβαζω, *send across.*
διαγιγνώσκω, *distinguish.*
διαδίδωμι, *distribute.*
διαδοχ-ή, *-ῆς*, f. *succession, turn.*
διαιρῶ[έω], *divide.*
διακηρυκεύομαι, *send a messenger.*
διακινδυνεύω, *risk everything, fight out the issue to the end.*
διακομίζω, *bring, convey across.*
διακρίνω, aor. pass. *διεκρίθην*, *separate.*
διαλλάσσω, aor. pass. *διηλλάγην*, *change, reconcile;* in pass. *become friends.*
διάλυσ-ις, *-εως*, f. *cessation.*
διαλύω, *terminate.*
διαμέλλω, *delay.*
διαμῶ[άω], *scrape through.*
διανοοῦμαι[έομαι], aor. *διενοήθην*, *intend.*
διαπλέω, *sail through* or *across.*
διαπλ-οῦς, *-οῦ*, m. *sailing through, passage.*
διαπράσσω, act. or mid., *carry through, accomplish.*
διασκευάζω, *arrange;* in mid. *make preparations.*
διασώζω, *save.*
διατάσσω, 3rd pl. plpf. mid. *διετετάχατο*, *post, arrange.*
διατρέφω, *sustain throughout.*
διαφέρω, aor. pass. *διηνέχθην*, *be different;* in mid. and pass. *be at variance.*
διαφεύγω, *escape.*
διαφθείρω, aor. *διέφθειρα*, aor. pass. *διεφθάρην*, fut. pass. *διαφθαρήσομαι*, *kill, destroy.*
διάφορ-ος, *-ον*, *important, different from* (with gen.).
διδάσκω, *teach, inform, advise.*
δίδωμι, *δώσω*, *ἔδωκα*, *give, offer.*
διίστημι, *separate;* in intr. tenses and pass. *be divided.*
δίκαι-ος, *-α*, *-ον*, *just;* **δίκαιός**

εἰμι, with inf. *be bound to, ought to.*
διότι, conj. *because.*
δίς, adv. *twice.*
δίψ-α, -ης, f. *thirst.*
διώκω, *pursue.*
δόκησ-ις, -εως, f. *reputation, opinion.*
δοκῶ[έω], δόξω, ἔδοξα, *seem, seem good, think;* impers. δοκεῖ, *it seems good.*
δόξ-α, -ης, f. *reputation, opinion.*
δοράτι-ον, -ου, n. *spear.*
δουλῶ[όω], *enslave, subdue.*
δρόμ-ος, -ου, m. *run.*
δρῶ[άω], ἔδρασα, *do.*
δύναμαι, δυνήσομαι, ἐδυνήθην, *be able.*
δύναμ-ις, -εως, f. *power, force.*
δυνατ-ός, -ή, -όν, *possible.*
δύο, gen. δυοῖν, *two.*
δυσέμβατ-ος, -ον, *difficult to move over;* **τὸ δυσέμβατον**, *ruggedness.*

ἔαρ, ἦρος, n. *spring.*
ἑαυτ-όν, -ήν, -ό, accus., reflexive pron. *himself, herself, itself;* in pl. *themselves.*
ἑβδομήκοντα, *seventy.*
ἐγγύς, adv. or prep., with gen. *near;* comp. ἐγγυτέρω, superl. ἐγγύτατα.
ἐγκαθέζομαι, *encamp in.*
ἐγκαθορμίζομαι, *come to anchor in.*
ἐγκαλῶ[έω], *accuse of.*
ἐγκαταλαμβάνω, aor. pass. ἐγκατελήφθην, *capture in, catch in, find in, bind by.*
ἐγκαταλείπω, *leave in.*
ἔγκειμαι, impf. ἐνεκείμην, *attack.*
ἔγκλημα, -τος, n. *ground of complaint, accusation.*
ἐγκύπτω, pf. ptcp. ἐγκεκυφώς, *stoop.*
ἐγχειρῶ[έω], *set to* (a task).
ἐγώ, ἐμοῦ or μου, *I;* pl. ἡμεῖς, ἡμῶν, *we.*
ἐθέλω, aor. ἠθέλησα, *be willing.*
ἔθ-ος, -ους, n. *custom.*
εἰ, conj. *if;* **εἰ καί**, *even if;* **εἴ πως** with opt., *in the hope that.*
εἰκάζω, aor. ᾔκασα, *compare, conjecture.*
εἴκοσι(ν), *twenty.*
εἰκότως, adv. of εἰκώς, *naturally.*
εἰκώς, see ἔοικα.
Εἵλ-ως, -ωτος, m. *Helot, serf* (of the Spartans).
εἰμί, ἔσομαι, impf. ἦν, *be;* impers. ἔστι, *it is impossible.*
εἶμι, impf. ᾖειν, inf. ἰέναι, *go.*
εἴπερ, *if.*
εἴργω, εἴρξω, εἶρξα, *hinder, bar the way, keep off.*
εἰρήν-η, -ης, f. *peace.*
εἷς, μία, ἕν, *one.*
εἴωθα, pf. of ἔθω with pres. meaning, *be accustomed;* ptcp. εἰωθώς, *customary;* **τὸ εἰωθός**, *custom.*
ἐκ (ἐξ before vowels), prep., with gen. *from, out of;* **ἐκ πολλοῦ**, *from a distance.*
ἕκαστ-ος, -η, -ον, *each.*
ἑκάτερ-ος, -α, -ον, *each* (of two), *either;* in pl. *each side, each army.*
ἑκατέρωθεν, adv. *on each side.*
ἑκατόν, *a hundred.*
ἐκβολ-ή, -ῆς, f. *putting forth, earing* (of corn).
ἐκδραμών, aor. ptcp. of ἐκτρέχω.

ἐκεῖν-ος, -η, -ο, dem. pron. *that, yonder, that man;* dat. fem. *ἐκείνῃ, there.*

ἐκεῖσε, adv. *thither.*

ἑκκαίδεκα, *sixteen.*

ἐκκρούω, *drive out.*

ἐκλογίζομαι, *reckon up.*

ἑκούσι-ος, *-ον, willing.*

ἐκπέμπω, *send out.*

ἐκπλέω, *sail out.*

ἔκπληξ-ις, *-εως*, f. *alarm, dismay.*

ἐκπλήσσω, aor. *ἐξέπληξα, dismay.*

ἐκπολιορκῶ[έω], *take by siege.*

ἐκτειχίζω, *fortify.*

ἐκτρέχω, aor. *ἐξέδραμον, run out, sally forth.*

ἐκφέρω, aor. pass. ptcp. *ἐξενεχθείς, carry out* or *ashore.*

ἐλάσσ-ων, *-ον, less, worse;* in pl. *fewer.*

ἐλάχιστ-ος, *-η, -ον, least;* in pl. *fewest.*

ἐλευθερί-α, *-ας*, f. *freedom.*

ἕλκω, aor. *εἵλκυσα, drag, draw.*

Ἕλλ-ην, *-ηνος*, m. *Greek.*

Ἑλληνικ-ός, *-ή, -όν, Greek;* **τὸ Ἑλληνικόν**, *the Greeks.*

ἐλπίζω, aor. *ἤλπισα, hope, expect.*

ἐλπ-ίς, *-ίδος*, f. *hope.*

ἐμβάλλω, *attack, ram.*

ἐμβοῶ[άω], *shout.*

ἐμμένω, *abide by* (with dat.).

ἐμπειρί-α, *-ας*, f. *experience, skill.*

ἔμπειρ-ος, *-ον, experienced in, acquainted with* (with gen.).

ἐμπίπρημι, aor. *ἐνέπρησα*, aor. pass. *ἐνεπρήσθην, burn.*

ἐμπίπτω, aor. *ἐνέπεσον, fall upon, attack* (with dat.)

ἐμφράσσω, aor. *ἐνέφαρξα, block.*

ἐν, prep., with dat. *in, at, among;* **ἐν τούτῳ**, *meanwhile.*

ἐνάγω, *urge on, influence.*

ἐνάντι-ος, *-α, -ον, opposite;* as noun, *enemy;* **ἐξ ἐναντίας**, *opposite.*

ἐναντιοῦμαι[όομαι], *oppose.*

ἐναποκλῶ[άω], plpf. mid. *ἐναπεκεκλάσμην, break off in.*

ἐνδε-ής, *-ές, insufficient;* comp. adv. *ἐνδεεστέρως, more sparingly.*

ἔνδει-α, *-ας*, f. *want.*

ἐνδέχομαι, *admit of, be possible.*

ἔνδηλ-ος, *-ον, clear, revealed.*

ἐνδίδωμι, *give in, make concessions.*

ἔνειμι, *be in.*

ἐννέα, *nine.*

ἐνταῦθα, adv. *here, then.*

ἐντεῦθεν, adv. *hence, thence.*

ἐντός, prep., with gen. *within.*

ἐντυγχάνω, *be* or *come in the way.*

ἕξ, *six.*

ἐξαγγέλλω, *bring a report.*

ἐξάγω, aor. *ἐξήγαγον, lead out.*

ἐξαναχωρῶ[έω], *withdraw from, evade.*

ἐξαπαλλάσσω, aor. pass. *ἐξαπηλλάγην, rid of.*

ἐξαπιναίως, adv. *suddenly.*

ἐξελαύνω, aor. *ἐξήλασα*, act. or mid. *drive out.*

ἐξεργάζομαι, *finish off.*

ἐξέρχομαι, *go out, take the field.*

ἔξεστι, impers. *it is possible.*

ἑξήκοντα, *sixty.*

ἐξίστημι, *put out, drive out;* in mid. *resign.*

ἐξουσί-α, *-ας*, f. *power, resources.*

ἔξω, adv. or prep., with gen. *outside.*

ἔοικα, pf. with pres. sense, *seem, seem likely;* ptcp. ἐοικώς or εἰκώς, *likely, natural;* **κατὰ τὸ εἰκός**, *probably;* **ἐκ τοῦ εἰκότος**, *naturally.*

ἑορτ-ή, -ῆς, f. *festival, feast;* **ἑορτὴν ἄγω**, *hold a festival.*

ἐπάγω, aor. ἐπήγαγον, *bring to, lead against;* in mid. *invite.*

ἐπαίρω, *raise;* in mid. *be elated.*

ἐπανέρχομαι, *come back.*

ἐπεί, conj. *when, since.*

ἐπείγω, impf. ἤπειγον, *urge on;* in mid. *hasten.*

ἐπειδή, conj. *when, since.*

ἔπειμι, *go against, attack* (with dat.), *follow.*

ἔπειτα, *then, next;* **τὸ ἔπειτα**, *the future.*

ἐπεκδρομ-ή, -ῆς, f. *sortie.*

ἐπεκθέω, *charge out against.*

ἐπεξέρχομαι, *follow up* (with dat.).

ἐπεπόνθεσαν, 3rd pl. plpf. of πάσχω.

ἐπέρχομαι, aor. **ἐπῆλθον**, *come to* or *upon, attack, approach* (with dat.).

ἐπερώτησ-ις, -εως, f. *question.*

ἐπεσβαίνω, *enter* (to meet the enemy).

ἐπεσπλέω, *sail in against.*

ἐπέχω, aor. ἐπεσχον, *check, wait.*

ἐπηρμένος, pf. ptcp. mid. of ἐπαίρω.

ἐπί, prep., with accus. *to, against, for;* with gen. *on, towards;* with dat. *at, on condition of, in the power of;* **ἐφ' ᾧ**, *on condition that.*

ἐπιβάλλω, ἐπιβαλῶ, ἐπέβαλον, aor. pass. ἐπεβλήθην, *cast on, fix to.*

ἐπιβιβάζω, *put on board.*

ἐπιβοηθῶ[έω], *attack, go to aid* (with dat.).

ἐπιβοῶ[άω], *cry out.*

ἐπιγίγνομαι, *come on, come after, follow.*

ἐπιδίδωμι, *give, sacrifice* (for).

ἐπιδιώκω, *pursue.*

ἐπιδρομ-ή, -ῆς, f. *attack.*

ἐπιεικ-ής, -ές, *fair, reasonable.*

ἐπιθέω, *charge, attack* (with dat.).

ἐπιθυμῶ[έω], *desire* (with gen.).

ἐπικαταβαίνω, *go down to meet* (the enemy).

ἐπίκειμαι, *lie over against, attack* (with dat.).

ἐπικελεύω, *urge, bid, exhort* (with dat.).

ἐπικηρυκεύομαι, *make overtures.*

ἐπικρατῶ[έω], *be victorious.*

ἐπιλαμβανω, *overtake, come upon;* in mid. *lay hold of* (with gen.).

ἐπίμαχ-ος, -ον, *open to attack.*

ἐπιμελοῦμαι[έομαι], aor. ἐπεμελήθην, *attend to, take in hand* (with gen.).

ἐπιμένω, *remain.*

ἐπινοῶ[έω], *plan.*

ἐπιπίπτω, *fall upon* (with dat.).

ἐπιπλέω, *be on board, sail against* (with dat.).

ἐπιπλ-οῦς, -οῦ, m. *attack* (of ships).

ἐπίπον-ος, -ον, *troublesome.*

ἐπιῤῥώννυμι, aor. ἐπέῤῥωσα, *encourage.*
ἐπισπέρχω, *urge on.*
ἐπισπόμενος, aor. ptcp. mid. of ἐφέπω, *pursue.*
ἐπίσταμαι, *know.*
ἐπιστέλλω, *send a message to, command* (with dat.); **τὰ ἐπεσταλμένα** (n. pl. pf. ptcp. mid.), *instructions.*
Ἐπιτάδ-ας, -ου, m. *Epitadas.*
ἐπιτήδει-ος, -α, -ον, *convenient, fit;* n. pl. as noun, *provisions.*
ἐπιτίθημι, *set on;* in mid. *attack* (with dat.).
ἐπιτιμῶ[άω], *blame, criticize* (with dat.).
ἐπιτρέχω, aor. ἐπέδραμον, *charge, attack* (with dat.).
ἐπιφέρω, *bring against;* **τὰ ὅπλα ἐπιφέρω**, *attack.*
ἐπιχείρησ-ις, -εως, f. *attempt, initiative.*
ἐπιχειρῶ[έω], *try, make an attempt.*
ἐπιχώρι-ος, -ον, *customary in one's country.*
ἐποκέλλω, *run aground.*
ἐργάζομαι, impf. εἰργαζόμην, *work, do;* pf. εἴργασμαι often in pass. sense.
ἔργ-ον, -ου, n. *work, action.*
ἐρημί-α, -ας, f. *deserted state.*
ἐρῆμ-ος, -ον, *desolate, deserted.*
ἔρομαι, aor. ἠρόμην, *ask.*
ἔρυμα, -τος, n. *fort.*
ἔρχομαι, aor. ἦλθον, *go.*
ἐς (εἰς), prep., with accus. *to, into.*
ἐσάγω, *bring to.*
ἐσακούω, *hear.*
ἐσβαίνω, aor. ἐσέβην, *go on board.*
ἐσβάλλω, aor. ἐσέβαλον, plpf. ἐσεβεβλήκειν, *attack, invade* (with dat.).
ἐσβιάζομαι, *force a way* (into).
ἔσειμι, *go into, occur to.*
ἐσνέω, *swim to.*
ἐσπέμπω, *send to* or *in.*
ἐσπίπτω, *fall* or *rush into.*
ἐσπλέω, *sail to* or *into.*
ἐσπλ-οῦς, -οῦ, m. *sailing in, entrance.*
ἔσχατ-ος, -η, -ον, *furthest;* **τὰ ἔσχατα**, *the edges.*
ἕτερ-ος, -α, -ον, *other, another;* **ὁ ἕτερος**, *the one, the other.*
ἔτι, adv. *still.*
ἑτοιμάζω, *make ready.*
ἑτοῖμ-ος, -ον, *ready, easily obtainable.*
εὔελπ-ις, -ι, *confident.*
εὐεργεσί-α, -ας, f. *favour, benefit.*
εὐθύς, adv. *immediately.*
εὐν-ή, -ῆς, f. *bed, couch.*
εὐξύνετ-ος, -ον, *shrewd, clever.*
εὐπορί-α, -ας, f. *abundance.*
εὔπορ-ος, -ον, *easy.*
εὐπραγί-α, -ας, f. *success, good fortune.*
εὑρίσκω, aor. εὗρον or ηὗρον, *find.*
Εὐρυμέδ-ων, -οντος, m. *Eurymedon.*
εὐρυχωρί-α, -ας, f. *open space, open sea.*
εὐτυχί-α, -ας, f. *good fortune.*
εὐτυχῶ[έω], *be fortunate.*
ἐφαιροῦμαι[έομαι], *choose as successor;* pf. ptcp. ἐφῃρημένος in pass. sense.
ἐφέλκω, *drag behind.*
ἐφέπω, aor. ἐπέσπον, act. and mid. *pursue.*

ἔφοδ-ος, *-ου*, f. *attack, march* (against), *approach*.
ἐφορμίζομαι, *come to anchor*.
ἔφορμ-ος, *-ου*, m. *blockading, anchorage*.
ἐφορμῶ[έω], *lie at anchor off, blockade* (with dat.).
ἔχθ-ος, *-ους*, n. *hate, enmity*.
ἔχθρ-α, *-ας*, f. *enmity*.
ἐχθρ-ός, *-ά*, *-όν*, *hostile*; as noun, *enemy*.
ἐχυρ-ός, *-ά*, *-όν*, *secure*.
ἔχω, *ἕξω* or *σχήσω*, *ἔσχον*, *have, keep*; with advs. *be*; in fut. and aor. often *put in* (at); in neg. followed by depend. clause, *know*.
ἐῶ[άω], aor. *εἴασα*, *allow*.
ἕως, *ἕω*, f. *dawn*.
ἕως, conj., with indic., *ἄν* and subjunct. or opt., *until*.

Ζάκυνθ-ος, *-ου*, f. *Zacynthus*.
ζῶ[άω], *live, be alive*.

ἤ, conj. *or*; **ἤ**...**ἤ**, *either*...*or*; after comp. *than*.
ᾗ, see *ὅς*.
ἡγοῦμαι[έομαι], *think, lead*.
ἤδη, adv. *already, now*.
ἡδον-ή, *-ῆς*, f. *pleasure*.
Ἠι-ών, *-όνος*, f. *Eion*.
ἥκιστα, superl. adv. *least*.
ἥκω, *have come*; fut. *ἥξω*, *will come*.
ἥλι-ος, *-ου*, m. *sun*.
ἡμεῖς, *ἡμῶν*, see *ἐγώ*.
ἡμέρ-α, *-ας*, f. *day*.
ἡμέτερ-ος, *-α*, *-ον*, *our, ours*.
ἡμισ-υς, *-εῖα*, *-υ*, *half*.
ἤν, conj. *if, in case* (with subjunct.).
ἤπειρ-ος, *-ου*, f. *mainland*.
ἠπειρώτ-ης, *-ου*, m. *landsman*.
ᾗπερ, see *ὅσπερ*.
ἥσσ-ων, *-ον*, *less, inferior*; adv. **ἧσσον**, *less*.
ἡσυχάζω, *be at leisure, be inactive, rest, stay still*.
ἡσυχί-α, *-ας*, f. *leisure*.

θαλαμι-ός, *-οῦ*, m. *rower on the lowest bench* of a trireme.
θάλασσ-α, *-ης*, f. *sea*.
θαλάσσι-ος, *-α*, *-ον*, *of the sea, nautical*.
θαρσῶ[έω], *take courage*.
θάσσ-ων, *-ον*, see *τάχιστος*.
Θεαγέν-ης, *-ους*, m. *Theagenes*.
θεράπ-ων, *-οντος*, m. *servant*.
Θερμοπύλ-αι, *-ων*, f. pl. *Thermopylae*.
θέρ-ος, *-ους*, n. *summer*.
θνήσκω, aor. *ἔθανον*, pf. ptcp. *τεθνεώς*, *die, be killed*.
θόρυβ-ος, *-ου*, m. *confusion, tumult*.
Θρᾴκ-η, *-ης*, f. *Thrace*.
Θρασυμηλίδ-ας, *-ου*, m. *Thrasymelidas*.

ἴδι-ος, *-α*, *-ον*, *private, peculiar, of one's own*.
ἰδιώτ-ης, *-ου*, m. *private person, one holding no official position*.
ἱκαν-ός, *-ή*, *-όν*, *sufficient*.
Ἴμβρι-οι, *-ων*, m. pl. *Imbrians*.
ἵνα, conj., with subjunct. or opt. *in order that*.
Ἱππαγρέτ-ας, *-ου*, m. *Hippagretas*.

ἰσθμ-ός, -οῦ, m. *isthmus.*
ἴσ-ος, -η, -ον, *equal;* **ἐν τῷ ἴσῳ**, *on equal terms.*
ἴστε, 2nd pers. pl. imperat. of **οἶδα**, *know.*
ἵστημι, *στήσω*, 1st aor. *ἔστησα*, 2nd aor. *ἔστην*, pf. *ἕστηκα*, *set up, set;* in intr. tenses and mid. *be set, stand.*
ἰσχυρίζομαι, *maintain, emphasize.*
ἰσχυρ-ός, -ά, -όν, *strong.*
ἰσχ-ύς, -ύος, f. *strength, power.*

καθάπερ, *as.*
καθίστημι, 1st aor. *κατέστησα*, 2nd aor. *κατέστην*, pf. *καθέστηκα*, *put, establish, appoint;* in intr. tenses and mid. *be established, made, appointed, become, stand, arise, come, prepare, arrange;* *καθεστηκώς*, *existing, present.*
καθορμίζω, *bring to anchor;* in mid. *put in.*
καθορῶ[άω], *see, observe.*
καί, conj. *and, also, even.*
καίπερ, *although* (with ptcp.).
καίρι-ος, -α, -ον, *opportune, favourable;* **τὰ καίρια**, *opportune circumstances.*
καιρ-ός, -οῦ, m. *occasion, chance.*
καίτοι, *and yet.*
καίω, pf. mid. *κέκαυμαι*, *burn.*
κακοπαθῶ[έω], *be in distress.*
κακ-ός, -ή, -όν, *bad;* n. sing. **κακόν** as noun, *evil, calamity;* comp. *κακίων*, superl. *κάκιστος.*
κακῶ[όω], *injure;* in mid. *suffer severely.*
καλ[έω], *καλῶ*, *ἐκάλεσα*, aor. pass. *ἐκλήθην*, *call.*
καλ-ός, -ή, -όν, *fair, good, honourable;* **καλὸς κἀγαθός**, *gallant;* comp. *καλλίων*, superl. *κάλλιστος.*
καλῴδι-ον, -ου, n. *small rope.*
κάλ-ως, -ω, m. *rope;* **ἀπὸ κάλω**, *in tow.*
Καμαρίν-η, -ης, f. *Camarina.*
καρτερ-ός, -ά, -όν, *strong.*
κατά, prep., with accus. *along, throughout, at, on the ground of, according to, in proportion to;* **κατὰ γῆν**, *by land;* **κατ' ὀλίγους**, *in small detachments;* **καθ' ὅτι**, *in what way;* **καθ' ἑαυτούς**, *by themselves;* **κατὰ τὸν ποταμόν**, *down the river.* With gen. *down from, at, against.*
καταβαίνω, *go down.*
κατάγνυμι, *break.*
καταδιώκω, *pursue, drive back.*
κατακαίω, aor. pass. *κατεκαύθην*, *burn down.*
καταλαμβάνω, aor. *κατέλαβον*, pf. mid. *κατείλημμαι*, *seize, come upon, overtake.*
καταλείπω, *leave.*
κατάλυσ-ις, -εως, f. *ending* (of war).
καταλύω, *break, destroy;* in mid. *make peace.*
καταπλέω, *sail down.*
καταπλήσσω, aor. pass. *κατεπλάγην*, *dismay.*
κατάπλ-ους, -ου, m. *sailing to land.*
καταπολεμῶ[έω], *subdue by attack.*

καταπροδίδωμι, *surrender, betray.*
κάταρσ-ις, -εως, f. *landing-place.*
κατάσκοπ-ος, -ου, m. *investigator.*
κατατίθημι, *set down;* in mid. *arrange.*
καταφαν-ής, -ές, *easily seen.*
καταφέρω, aor. κατήνεγκον, *carry down, drive to land.*
καταφεύγω, *flee for refuge.*
καταφρονῶ[έω], *despise.*
κατέχω, κατασχήσω, *gain control of, check, be stationed.*
κατόπιν, adv. *behind.*
κόχλ-ηξ, -ηκος, m. *shingle.*
κεῖμαι, impf. ἐκείμην, *lie, be placed.*
κέκαυμαι, pf. mid. of καίω, *burn.*
κελεύω, *bid, advise.*
κέλ-ης, -ητος, m *cutter.*
κεν-ός, -ή, -όν, *empty, not manned.*
Κερκυραῖ-ος, -α, -ον, *Corcyraean.*
κήρυγμα, -τος, n. *proposal, proclamation.*
κῆρ-υξ, -υκος, m. *envoy, herald.*
κηρύσσω, aor. ἐκήρυξα, pf. mid. κεκήρυγμαι, *invite, proclaim, propose.*
κινδυνεύω, *be in danger, run a risk.*
κίνδυν-ος, -ου, m. *danger, risk.*
Κλέ-ων, -ωνος, m. *Cleon.*
κληθείς, aor. ptcp. pass. of καλῶ.
κλῄω, *close, block.*
κοιν-ός, -ή, -όν, *common* (to all).
κοινῶ[όω], *communicate.*
κολυμβητ-ής, -οῦ, m. *diver.*
κομιδ-ή, -ῆς, f. *conveyance.*
κομίζω, κομιῶ, ἐκόμισα, *bring, convey;* in mid. *recover.*
κονιορτ-ός, -οῦ, m. *dust.*

κόπτω, pf. mid. κέκομμαι, *strike, disable, pound, grind.*
Κορυφάσι-ον, -ου, n. *Coryphasium.*
κόσμ-ος, -ου, m. *honour, credit.*
κοτύλ-η, -ης, f. *about half a pint.*
κοῦφ-ος, -η, -ον, *light.*
Κρατησικλ-ῆς, -έους, m. *Cratesicles.*
κράτιστ-ος, -η, -ον, *best, strongest.*
κράτ-ος, -ους, n. *power;* **κατὰ κράτος**, *with all one's might.*
κρατῶ[έω], *overcome, gain possession of* (with gen.), *hold, gain the upper hand.*
κρέ-ας, -ως, n. *meat.*
κρείσσ-ων, -ον, *better, superior;* **τὰ κρείσσω**, *advantages.*
κρημνώδ-ης, -ες, *precipitous;* **τὸ κρημνῶδες**, *the cliff.*
κρήν-η, -ης, f. *spring, well.*
κυβερνήτ-ης, -ου, m. *pilot.*
κυκλῶ[όω], *encircle.*
κύκλωσ-ις, -εως, f. *encircling.*
κύρι-ος, -α, -ον, *having power, having the deciding voice* (with gen.).
κωλύμ-η, -ης, f. *hindrance.*
κωλύω, *hinder, prevent.*

λάθρᾳ, adv. *secretly.*
Λακεδαιμόνι-ος, -α, -ον, *Lacedaemonian.*
Λακωνικ-ός, -ή, -όν, *Laconian;* **ἡ Λακωνική**, *Laconia.*
λαμβάνω, λήψομαι, ἔλαβον, aor. pass. ἐλήφθην, *take, receive.*
λανθάνω, λήσω, ἔλαθον, *escape the notice of;* in ptcp. *unobserved.*

λέγω, *ἐρῶ*, *εἶπον* or *ἔλεξα*, *εἴρηκα*, *say*, *tell*, *bid*, *mean*.

Λεοντῖν-οι, *-ων*, m. pl. *Leontines*.

Λευκάδι-οι, *-ων*, m. pl. *Leucadians*.

λῄζομαι, *make raids on*.

Λήμνι-οι, *-ων*, m. pl. *Lemnians*.

λῃστεί-α, *-ας*, f. *raiding*.

λῃστεύω, *plunder*, *raid*.

λῃστρικ-ός, *-ή*, *-όν*, *of a pirate* or *pirates*; **λῃστρική (ναῦς)**, *pirate-ship*.

λίθ-ος, *-ου*, m. *stone*.

λιθουργ-ός, *-όν*, *for stone-breaking*.

λιμ-ήν, *-ένος*, m. *harbour*.

λιμ-ός, *-οῦ*, m. *famine*, *hunger*.

λίν-ον, *-ου*, n. *lint*.

λιποψυχῶ[έω], *swoon*.

λογάδην, adv. *as picked up*.

λογίζομαι, *reflect*.

λογισμ-ός, *-οῦ*, m. *reckoning*, *reflection*.

λόγ-ος, *-ου*, m. *word*, *proposal*, *reckoning*; **ἐς λόγους ξυνέρχομαι**, *meet in conference*.

λοιπ-ός, *-ή*, *-όν*, *remaining*, *rest*.

Λοκρ-ίς, *-ίδος*, f. adj. *Locrian*.

Λοκρ-οί, *-ῶν*, m. pl. *Locrians*.

λόχ-ος, *-ου*, m. *company* (of soldiers).

λύω, *loose*, *terminate*.

μακρ-ός, *-ά*, *-όν*, *long*; **μακρὰ ναῦς**, *war-ship*; **ἐπὶ μακρότερον**, *on a larger scale*.

μάλα, adv. *very*, *very much*; comp. *μᾶλλον*, *more*, *rather*; superl. *μάλιστα*, *most*, *chiefly*, *about* (of numbers).

μανιώδ-ης, *-ες*, *seeming like madness*.

μάρτ-υς, *-υρος*, m. *witness*.

μάσσω, pf. ptcp. mid. *μεμαγμένος*, *knead*.

μάχ-η, *-ης*, f. *fight*, *fighting*.

μάχομαι, *μαχοῦμαι*, *ἐμαχεσάμην*, *fight*, *fight against* (with dat.).

μέγ-ας, *-άλη*, *-α*, *great*; comp. *μείζων*, superl. *μέγιστος*; comp. adv. *μειζόνως*, *to a greater degree*.

μέγεθ-ος, *-ους*, n. *size*.

μελιτῶ[όω], *mix with honey*.

μέλλω, *μελλήσω*, *be about to*, *be likely to*, *intend*, *be intended to*, *delay*.

μεμαγμένος, see *μάσσω*.

μέν, part. *on the one hand*, contrasting one clause or sentence with a second containing *δέ*.

Μενδαῖ-οι, *-ων*, m. pl. *Mendaeans*.

μένω, *μενῶ*, *ἔμεινα*, *remain*, *stand one's ground*.

μέρ-ος, *-ους*, n. *part*, *turn*.

μέσ-ος, *-η*, *-ον*, *middle*.

Μεσσήν-η, *-ης*, f. *Messana*.

Μεσσήνι-ος, *-α*, *-ον*, *Messenian* or *Messanian*.

Μεσσην-ίς, *-ίδος*, f. adj. *Messenian*.

μετά, prep., with accus. *after*; with gen. *with*.

μεταβολ-ή, *-ῆς*, f. *change*.

μεταμέλομαι, *repent*.

μεταξύ, adv. or prep., with gen. *between*; **τὸ μεταξύ**, *the space between*.

μεταπέμπω, act. or mid. *send for*.

μεταχειρίζω, *take in hand*.

μετέωρ-ος, *-ον*, *out at sea, high;* **τὸ μετέωρον**, *high ground.*
μέτρι-ος, *-α, -ον*, *moderate, comparatively small, not strict;* adv. *μετρίως*, *with moderation, on moderate terms.*
μέχρι, prep., with gen. or conj. *until;* also *μέχρι οὗ* with indic. or subjunctive.
μή, neg. part. in commands and final, conditional, and generic clauses, *not;* after verbs of fearing *lest.*
μηδ-είς, *-εμία, -έν*, *no one, no....*
μηκέτι, adv. *no longer.*
μηκυνω, *μηκυνῶ*, *lengthen, prolong.*
μήκ-ων, *-ωνος*, f. *poppy, poppy-seed.*
μηχαν-ή, *-ῆς*, f. *engine of war.*
μικρ-ός, *-ά, -όν*, *little, small.*
μόλις, adv. *hardly, with difficulty.*
Μόλοβρ-ος, *-ου*, m. *Molobrus.*

Νάξ-ος, *-ου*, f. *Naxus.*
ναυάγι-ον, *-ου*, n. *wreck.*
ναύαρχ-ος, *-ου*, m. *commander* (of ships).
ναυμαχί-α, *-ας*, f. *sea-fight.*
ναυμαχῶ[έω], *fight a naval battle.*
Ναύπακτ-ος, *-ου*, f. *Naupactus.*
ναῦς, *νεώς*, pl. *νῆες*, f. *ship.*
ναύτ-ης, *-ου*, m. *sailor.*
ναυτικ-ός, *-ή, -όν*, *of* or *from ships;* **τὸ ναυτικόν**, *fleet.*
νεκρ-ός, *-οῦ*, m. *dead body.*
νεωστί, adv. *lately.*
νεωτερίζω, *revolutionize.*
νῆσ-ος, *-ου*, f. *island.*
Νικήρατ-ος, *-ου*, m. *Niceratus.*
Νικί-ας, *-ου*, m. *Nicias.*
νικῶ[άω], *defeat, surpass, conquer, win.*
Νίσαι-α, *-ας*, f. *Nisaea.*
νομίζω, *νομιῶ, ἐνόμισα*, *think.*
νόμ-ος, *-ου*, m. *law, custom.*
νοῦς, *νοῦ*, m. *mind;* **ἐν νῷ ἔχω**, *intend.*
νύξ, *νυκτός*, f. *night.*
νῦν, adv. *now.*
νῶτ-ον, *-ου*, n. *back;* **κατὰ νώτου**, *at one's back, in the rear.*

ξύγκειμαι, *be agreed.*
ξυγκλῄω, *close up.*
ξυγχωρῶ[έω], *agree, make concessions.*
ξυλλέγω, aor. *ξυνέλεξα*, aor. pass. *ξυνελέγην*, *collect;* in pass. *meet.*
ξύλ-ον, *-ου*, n. *wood, timber.*
ξυμβαίνω, aor. *ξυνέβην*, pf. *ξυμβέβηκα*, aor. subjunct. pass. (3rd sing.) *ξυμβαθῇ*, *happen, fit, come together, make an agreement.*
ξύμβασ-ις, *-εως*, f. *treaty.*
ξυμμαχί-α, *-ας*, f. *alliance.*
ξύμμαχ-ος, *-ον*, *fighting on the side of;* as noun, *ally.*
ξύμ-πας, *-πασα, -παν*, *whole.*
ξυμπλέκω, *join, clasp together.*
ξύμπτωμα, *-τος*, n. *plight, misfortune.*
ξυμφέρω, *be convenient, useful.*
ξυμφορ-ά, *-ᾶς*, f. *disaster.*
ξυναίρω, aor. *ξυνῆρα*, in mid. *take part in, share* (with gen.).

ξυναλλαγ-ή, *-ῆς*, f. *making of peace.*

ξυναλλάσσω, aor. pass. *ξυνηλλάγην*, *bring together;* in mid. and pass. *make peace.*

ξύνεγγυς, adv. *near together.*

ξύνεδρ-ος, *-ου*, m. *delegate* (*to sit in council with*, followed by dat.).

ξυνεθίζω, pf. mid. *ξυνείθισμαι*, *accustom;* in mid. *grow accustomed.*

ξύνειμι, *be with, come home to.*

ξυνεκπλέω, *join a (naval) expedition.*

ξυνεπάγω, *join in calling in.*

ξυνέρχομαι, aor. *ξυνῆλθον*, *meet.*

ξυνετ-ός, *-ή*, *-όν*, *clever.*

ξυντάσσω, aor. *ξυνέταξα*, *draw up.*

ξυντίθημι, aor. mid. *ξυνεθέμην*, *set together;* in mid. *agree to.*

ξυντρίβω, aor. *ξυνέτριψα*, *crush.*

ὁ, *ἡ*, *τό*, def. article, *the;* dem. pron. *he;* **ὁ δέ**, *but he;* dat. fem. *τῇ* as adv. *here, there;* **οἱ μέν...οἱ δέ**, *some...others, one side ..the other side.*

ὅδε, *ἥδε*, *τόδε*, dem. pron. or adj. *this.*

ὁδ-ός, *-οῦ*, f. *road.*

Ὀδυσσ-εύς, *-έως*, m. *Odysseus.*

ὅθεν, adv. *whence, from where, from which.*

οἱ, dat. of reflex. pron. *ἕ*, *to him.*

οἶδα, *εἴσομαι*, *ᾔδη*, ptcp. *εἰδώς*, *know.*

οἰκεῖ-ος, *-α*, *-ον*, *familiar, closely affecting, of one's own.*

οἰκειότ-ης, *-ητος*, f. *cordial relations.*

οἰκοδόμημα, *-τος*, n. *building.*

οἶκ-ος, *-ου*, m. *house, home.*

οἶν-ος, *-ου*, m. *wine.*

οἴομαι, aor. *ᾠήθην*, *think.*

οἷ-ος, *-α*, *-ον*, *such as;* **οἷός τέ εἰμι**, *be able, possible;* neut. *οἷον* as adv. *as.*

οἷόσπερ, *οἵαπερ*, *οἷόνπερ*, adj. *just as.*

οἰστ-ός, *-οῦ*, m. *arrow.*

οἰσύιν-ος, *-η*, *-ον*, *of wickerwork.*

ὀκέλλω, *run aground.*

ὀκτακόσι-οι, *-αι*, *-α*, *eight hundred.*

ὀκτώ, *eight.*

ὀλίγ-ος, *-η*, *-ον*, *small;* in pl. *few;* neut. *ὀλίγον* as adv. *a little;* **κατ' ὀλίγον**, *few at a time;* comp. *ἐλάσσων*, superl. *ἐλάχιστος.*

ὀλιγωρί-α, *-ας*, f. *contempt.*

ὁμαλ-ός, *-ή*, *-όν*, *level.*

ὅμοι-ος, *-α*, *-ον*, *like equal;* **ἐκ τοῦ ὁμοίου**, *on equal terms.*

ὅμορ-ος, *-ον*, *neighbouring.*

ὁμόσε, adv. *to meet.*

ὁμόφων-ος, *-ον*, *speaking the same tongue.*

ὅμως, conj. *nevertheless.*

ὀξ-ύς, *-εῖα*, *-ύ*, *sharp, quick.*

ὅπῃ, adv. *where.*

ὀπίσω, adv. *back;* **εἰς τοὐπίσω**, *behind.*

ὁπλίζω, aor. *ὥπλισα*, *arm.*

ὁπλίτ-ης, *-ου*, m. *heavy-armed foot-soldier, hoplite.*

ὅπλ-ον, -ου, n., in pl. *arms, armour*.

ὁπόθεν, adv. *from where*.

ὁπόσ-ος, -η, -ον, *how great, as great, as much as*.

ὁποσοσοῦν, neut. ὁποσονοῦν, *however much, however little*.

ὁπόταν, conj. *whenever*.

ὁπότε, conj. *when*.

ὁπότερ-ος, -α, -ον, *which* (of two); in pl. *which of two sides* or *armies*.

ὅπως, adv. *how, as;* conj. *in order that* (with subjunct. or opt.).

ὀρέγω, *stretch out;* in mid. *desire, grasp at* (with gen.).

ὀρθῶ[όω], *set straight, set right;* in mid. or pass. *succeed;* **τὸ ὀρθούμενον**, *success*.

ὅρκ-ος, -ου, m. *oath*.

ὁρμ-ή, -ῆς, f. *impulse*.

ὅρμ-ος, -ου, m. *anchorage*.

ὁρμῶ[άω], *set in motion;* intr. *start*, (with ἐπί) *attack;* in mid. *start, rush, be eager*.

ὁρμῶ[έω], *anchor*.

ὄρ-ος, -ους, n. *mountain*.

ὁρῶ[άω], ὄψομαι, εἶδον, ἑώρακα, *see*.

ὅς, ἥ, ὅ, relat. pron. *who, which;* dat. f. ᾗ, *for which reason, as, where;* **ἔστιν ᾗ**, *in some places;* gen. οὗ, *where*. Also demonstr. pron. after καί, *he, she, it*.

ὅσ-ος, -η, -ον, *how great, as great as;* in pl. *how many, as many as*.

ὅσπερ, ἥπερ, ὅπερ, relat. pron. *who, which;* dat. f. **ᾗπερ**, *where*.

ὅστις, ἥτις, ὅτι, *who, which, whoever, whatever*.

ὁστισοῦν, ὁτιοῦν, *any at all*.

ὅταν, conj. *whenever*.

ὅτε, conj. *when*.

ὅτι, conj. *that, because*.

ὅτι, neut. of ὅστις, used with superl. *as...as possible*, e.g. ὅτι τάχιστα, *as quickly as possible*.

ὅτιπερ, conj. *because*.

οὐ, οὐκ, οὐχ, neg. part. *not*.

οὗ, see ὅς.

οὐδέ, neg. part. *nor, and not, not even*.

οὐδ-είς, -εμία, -έν, *no one, no;* neut. sing. as adv. *not at all*.

οὐκέτι, adv. *no longer*.

οὖν, adv. *therefore*.

οὔτε...οὔτε, *neither...nor*.

οὗτος, αὕτη, τοῦτο, dem. pron. or adj. *this;* dat. f. ταύτῃ as adv. *here*.

οὕτω(ς), adv. *thus, so*.

ὀφείλω, *owe, be obliged*.

ὄχλ-ος, -ου, m. *crowd*.

ὀψέ, adv. *late*.

ὄψ-ις, -εως, f. *sight*.

πάθ-ος, -ους, n. *disaster*.

παλαι-ός, -ά, -όν, *ancient, old*.

πάλιν, adv. *back, again*.

πανδημεί, adv. *with the whole people, en masse*.

πανστρατιᾷ, *with their whole force*.

πανταχόθεν, adv. *from all sides*.

παρά, prep., with accus. *along*,

to, contrary to, beyond; with gen. *from*; with dat. *with, among*.

παραβαίνω, 3rd sing. aor. subjunct. pass. *παραβαθῇ*, *transgress, break* (a law, agreement etc.).

παραβοηθῶ[έω], *come to the rescue*.

παραγγέλλω, *give a command*.

παραγίγνομαι, *be near, come up*.

παραδίδωμι, *give up*.

παραδωσείω, *be willing to give up*.

παραινῶ[έω], *advise*.

παρακελεύομαι, *utter an exhortation*.

παρακελευσμ-ός, -οῦ, m. *cry of encouragement*.

παρακινδυνεύω, *run a risk*.

παρακομίζω, *bring along*; in mid. or pass. *sail along*.

παραλαμβάνω, *take over, receive*.

παραπέμπω, *send along*.

παραπίπτω, *occur*.

παραπλέω, plpf. *παρεπεπλεύκειν*, *sail along the coast*.

παράπλ-ους, -οῦ, m. *coasting voyage*.

παρασκευάζω, act. or mid. *prepare*.

παρασκευ-ή, -ῆς, f. *preparation, armament, force*.

παράσπονδ-ος, -ον, *contrary to an agreement*.

παρατάσσω, pf. mid. *παρατέταγμαι*, *draw up*.

παρατείνω, *stretch along*.

παρατυγχάνω, *be offered by chance*.

παραχρῆμα, adv. *immediately*.

παρείκω, *yield, afford a footing*.

πάρειμι, impf. *παρῆν*, *be present*; **ἐν τῷ παρόντι**, *at the present time*; impers. *πάρεστι*, *it is possible*.

παρεξειρεσί-α, -ας, f. *part of the ship not occupied by rowers, bows*.

παρέρχομαι, *come forward*.

παρέχω, *afford, inflict, cause*.

παρίημι, aor. *παρῆκα*, *let pass, lower*.

πᾶς, *πᾶσα*, *πᾶν*, *all, every*.

πάσχω, *πείσομαι*, *ἔπαθον*, *πέπονθα*, *suffer, be treated, fare*.

πατρ-ίς, -ίδος, f. *one's own country*.

παύω, *stop, make to cease*; in mid. *cease*.

πεζομαχῶ[έω], *fight a land battle*.

πεζ-ός, -ή, -όν, *on foot, on land, military*; **ὁ πεζός**, *the army, the foot-soldiers*; **πεζῇ**, *by land*.

πείθω, *πείσω*, *ἔπεισα* or *ἔπιθον*, *persuade*; in mid. *obey, be persuaded*.

πειρῶ[άω], act. or mid. *try, assault*.

πέλαγ-ος, -ους, n. *open sea, sea*.

Πελοποννήσι-ος, -α, -ον, *Peloponnesian*.

Πελοπόννησ-ος, -ου, f. *Peloponnese*.

πελταστ-ής, -οῦ, m. *targeteer, light-armed soldier*.

Πελωρ-ίς, -ίδος, f. *Peloris*.

πέμπω, *send*.

πέντε, *five*.

περί, prep., with accus. *about, around, with* (a person); with gen. *about, concerning*.

περιαγγέλλω, *send a message round*.

περιαλγῶ[έω], *be greatly grieved* or *dismayed*.

περιγίγνομαι, *prevail, survive, escape*.

περίειμι, inf. περιεῖναι, *be left, survive*.

περίειμι, inf. περιιέναι, *go round*.

περιέρχομαι, *go round*.

περιίστημι, 2nd aor. περιέστην, pf. ptcp. περιεστώς, *set round*; intr. and mid. *stand round, surround, come* or *change round*.

περίοδ-ος, -ου, f. *way round, surrounding*.

περίοικ-ος, -ου, m. *provincial* (of Laconia).

περιορμῶ[έω], *anchor round*.

περιορῶ[άω], *overlook, suffer, allow* (with accus. and ptcp. or inf.).

περιπλέω, *sail round*.

περίπλε-ως, -ων, *full*.

περιῤῥέω, aor. περιεῤῥύην, *slip off*.

Πέρσ-ης, -ου, m. *Persian*.

πετρώδ-ης, -ες, *rocky*.

πέφυκα, see φύω.

Πηγ-αί, -ῶν, f. pl. *Pegae*.

πηλ-ός, -οῦ, m. *clay, mortar*.

πιέζω, aor. ἐπίεσα, *distress, bring great suffering on*.

πιθαν-ός, -ή, -όν, *persuasive, influential*.

πῖλ-ος, -ου, m. *felt cap* or *cuirass*.

πίνω, πίομαι, ἔπιον, *drink*.

πίπτω, πεσοῦμαι, ἔπεσον, *fall*.

πιστεύω, *believe, trust in* (with dat.).

πλάγι-ος, -α, -ον, *sideways, on the flanks*; **ἐκ πλαγίου**, *on the flanks*.

πλέω, πλεύσομαι, ἔπλευσα, *sail*.

πλέων, see πολύς.

πλῆθ-ος, -ους, n. *number, large number, multitude*.

πλήν, adv. or prep., with gen. *except*.

πληρῶ[όω], *fill, man* (a ship).

πλοῖ-ον, -ου, n. *boat, ship*.

πλ-οῦς, -οῦ, m. *voyage*.

πνεῦμα, -τος, n. *breeze*.

ποιῶ[έω], *make, do*; in mid. *make, carry on, hold, esteem*.

πολέμι-ος, -α, -ον, *hostile, at war with*; as noun, *enemy*; adv. πολεμίως, *as enemies*.

πόλεμ-ος, -ου, m. *war*.

πολεμῶ[έω], *make war, fight*.

πολεμῶ[όω], *make an enemy, drive to war*.

πολιορκί-α, -ας, f. *siege, blockade*.

πολιορκῶ[έω], *besiege, blockade*.

πόλ-ις, -εως, f. *city, state*.

πολλάκις, adv. *often*.

πολλαπλάσι-ος, -α, -ον, *many times as much* or *as many*.

πολλαχόθεν, *from many sides, for many reasons*.

πολύς, πολλή, πολύ, *much, great, long*; in pl. *many*; neut. pl. as adv. *often*; comp. πλείων or πλέων, *more*; superl. πλεῖστος, *most*; **οἱ πολλοί** or **οἱ πλείονες**, *most* (men); **ὡς ἐπὶ πλεῖστον** *as much as possible.*

πόν-ος, -ου, m. *labour, harm, distress.*

πόντ-ος, -ου, m. *open sea.*

πονῶ[έω], *labour.*

πορθμ-ός, -οῦ, m. *straits.*

ποτε, indef. part. *at some time, once, ever.*

που, indef. adv. *somewhere, anywhere.*

πορίζω, *πoριῶ, ἐπόρισα, provide.*

ποταμ-ός, -οῦ, m. *river.*

πρᾶγμα, -τος, n. *thing, matter, affair;* in pl. often *position, fortunes.*

πράσσω, *πράξω, ἔπραξα, do, act, fare, bring to pass, negotiate.*

πρεσβεύω, *act as ambassador;* in mid. *send an embassy.*

πρέσβ-υς, -εως, m. *old man;* in pl. *ambassadors.*

πρίν, conj. or adv. *before;* as conj. followed by inf. or *ἄν* with subjunctive; **ἐν τῷ πρίν**, *before, formerly.*

πρό, prep., with gen. *before.*

προαφικνοῦμαι[έομαι], plpf. *προαφίγμην, arrive beforehand.*

προδίδωμι, *betray.*

προεμβάλλω, *ram first.*

προθυμί-α, -ας, f. *energy, eagerness.*

προθυμοῦμαι[έομαι], *be eager, make an effort.*

προκαλοῦμαι[έομαι], *invite, propose.*

προλαμβάνω, *get the advantage* (*in*, with gen.).

προλέγω, *say* or *order beforehand.*

προορῶ[άω], *see in front.*

προπέμπω, *send on in front.*

πρός, prep., with accus. *to, towards, in view of, with;* with gen. *in favour of, on the side of;* with dat. *in addition to.*

προσαιροῦμαι[έομαι], *choose in addition.*

προσβαίνω, *approach.*

προσβοηθῶ[έω], *come also to aid.*

προσβολ-ή, -ῆς, f. *attack, base of operations.*

προσγίγνομαι, *be added to, be acquired.*

προσδέομαι, *require* (with gen.).

προσδέχομαι, *expect, accept.*

προσδοκί-α, -ας, f. *expectation.*

πρόσειμι, inf. *προσεῖναι, be added to, be there, belong to.*

πρόσειμι, inf. *προσιέναι, approach* (with dat.).

προσέρχομαι, aor. *προσῆλθον, approach* (with dat.).

προσέχω, *come to land, put in.*

πρόσθεν, adv. or prep., with gen. *in front* (*of*).

προσίημι, *send to;* in mid. *accept.*

πρόσκειμαι, *press hard on, pursue* (with dat.).

προσκομίζω, *bring to, convey to.*

προσλαμβάνω, *gain,* or *take, in addition.*

προσμίγνυμι, aor. *προσέμιξα, approach, engage with* (with dat.).

προσόρμισ-ις, -εως, f. *landing-place.*

πρόσοψ-ις, -εως, f. *sight, power to see.*

προσπίπτω, *fall upon, attack* (with dat.).
προσπλέω, *sail towards*.
προσταυρῶ[όω], *defend*, or *join, by a palisade*.
προστίθημι, *attach, give*.
προσφέρω, *bring to;* in mid. *meet* (with dat.).
πρότερ-ος, -*α*, -*ον*, *former;* neut. sing. as adv. *before, first;* superl. *πρῶτος*, *first;* neut. sing. as adv. *first*.
προὔργου [for *πρὸ ἔργου*], *useful*.
προὔχω [*προέχω*], *excel*.
προφυλακ-ή, -*ῆς*, f. *advanced guard, outpost*.
πρόχειρ-ος, -*ον*, *ready to hand*.
προχωρῶ[έω], *go well, succeed*.
πρῴ, adv. *early*.
Πρωτ-ή, -*ῆς*, f. *Prote*.
πρῶτος, see *πρότερος*.
πταίω, *come to grief*.
Πυθόδωρ-ος, -*ου*, m. *Pythodorus*.
Πύλ-ος, -*ου*, f. *Pylus*.
πυνθάνομαι, *πεύσομαι*, *ἐπυθόμην*, *learn*.

ῥάδι-ος, -*α*, -*ον*, *easy;* comp. *ῥᾴων*, superl. *ῥᾷστος;* **ῥαδίως φέρω**, *take lightly*.
ῥαχί-α, -*ας*, f. *breakers*.
Ῥηγῖν-οι, -*ων*, m. pl. *Rhegines, men of Rhegium*.
ῥόθι-ον, -*ου*, n. *surge*.
ῥοώδ-ης, -*ες*, *with strong currents*.
ῥώμ-η, -*ης*, f. *power, confidence*.

σαφ-ής, -*ές*, *clear*.
σιδήρι-ον, -*ου*, n. *iron tool*.
σιδηρ-οῦς, -*ᾶ*, -*οῦν*, *of iron*.
Σικελί-α, -*ας*, f. *Sicily*.
Σικελ-οί, -*ῶν*, m. pl. *Sicels*.
Σιμωνίδ-ης, -*ου*, m. *Simonides*.
σίτι-ον, -*ου*, n. *food*.
σιτοδεί-α, -*ας*, f. *want of food*.
σιτοδοτῶ[έω], *supply with provisions*.
σῖτ-ος, -*ου*, m. *corn, food*.
σκευάζω, pf. mid. *ἐσκεύασμαι*, *equip, accoutre*.
σκοπῶ[έω], *consider, watch for*.
σμικρ-ός, -*ά*, -*όν*, *small*.
Σοφοκλ-ῆς, -*έους*, m. *Sophocles*.
σπανίζω, *lack* (with gen.).
Σπάρτ-η, -*ης*, f. *Sparta*.
Σπαρτιάτ-ης, -*ου*, m. *Spartiate*.
σπένδω, pf. inf. mid. *ἐσπεῖσθαι*, *pour a libation;* in mid. *make a treaty*.
σπέρμα, -*τος*, n. *seed;* **λίνου σπέρμα**, *linseed*.
σπονδ-ή, -*ῆς*, f. *libation;* in pl. *truce, treaty*.
σπουδ-ή, -*ῆς*, f. *energy, earnestness*.
στάδι-ον, -*ου*, n., pl. *στάδι-οι* or -*α*, *stade*, i.e. a distance of 606 ft.
στάδι-ος, -*α*, -*ον*, *close, hand to hand*.
στασιάζω, *be in a state of faction*.
στέγω, *keep out, be a protection against*.
στεν-ός, -*ή*, -*όν*, *narrow*.
στενότ-ης, -*ητος*, f. *narrowness*.
στενοχωρί-α, -*ας*, f. *want of room*.
στερῶ[έω], *deprive*.

πορίζω, *ποριῶ, ἐπόρισα, provide.*

στρατεί-α, -ας, f. *campaign*.
στράτευμα, -τος, n. *army*.
στρατεύω, *march, take the field.*
στρατηγ-ός, -οῦ, m. *commander*.
στρατηγῶ[έω], *be commander*.
στρατιώτ-ης, -ου, m. *soldier*.
στρατοπεδεύομαι, *encamp, take up quarters.*
στρατόπεδ-ον, -ου, n. *camp, naval station, army.*
στρατ-ός, -οῦ, m. *army*.
Στύφ-ων, -ωνος, m. *Styphon*.
σύ, σοῦ or σου, pl. ὑμεῖς, ὑμῶν, *you*.
Συρακόσι-ος, -α, -ον, *Syracusan*.
Σφακτηρί-α, -ας, f. *Sphacteria*.
σφάλλω, σφαλῶ, ἔσφηλα, aor. pass. ἐσφάλην, *cause to fail, deceive;* in pass. *suffer a calamity, be mistaken.*
σφᾶς (accus. of σφεῖς), gen. σφῶν, dat. σφίσι, reflex. pron. *them*.
σφενδόν-η, -ης, f. *sling*.
σφέτερ-ος, -α, -ον, *his own, their own.*
σχολάζω, *be at leisure*.
σώζω, σωσω, ἔσωσα, *save*.
σῶμα, -τος, n. *body, person*.
σώφρ-ων, -ον, *prudent, sober-minded.*

τακτ-ός, -ή, -όν, *fixed*.
ταλαιπωρῶ[έω], act. or pass. *be distressed, suffer hardships.*
ταξίαρχ-ος, -ου, m. *tribe-commander, subordinate officer.*
ταράσσω, pf. mid. τετάραγμαι, *confuse*.
τάσσω, aor. ἔταξα, pf. mid. τέταγμαι, *draw up, post, fix a price.*
ταύτῃ, see **οὗτος**.
τάχ-ος, -ους, n. *speed;* **κατὰ ταχος**, *with speed.*
ταχ-ύς, -εῖα, -ύ, *swift, quick;* comp. θάσσων, superl. τάχιστος; **διὰ ταχέων**, *hastily*.
τε, followed by καί or τε, *both ...and.*
τεθνεώς, pf. ptcp. of θνήσκω, *die*.
τειχήρ-ης, -ες, *within walls, besieged.*
τειχίζω, *fortify*.
τείχισμα, -τος, n. *wall, fortification.*
τεῖχ-ος, -ους, n. *wall, fortification.*
τελευταῖ-ος, -α, -ον, *last*.
τέλ-ος, -ους, n. *end;* as adv. *at last;* in pl. *magistrates.*
τεσσαράκοντα, *forty*.
τετρακόσι-οι, -αι, -α, *four hundred.*
τεχνῶμαι[άομαι], *devise means.*
τηρῶ[έω], *watch, wait, wait for, keep under guard.*
τίθημι, θήσω, ἔθηκα, aor. mid. ἐθέμην, *set, place;* in mid. *arrange.*
τιμ-ή, -ῆς, f. *honour*.
τιμῶ[άω], *honour, value.*
τιμωρί-α, -ας, f. *help, punishment.*
τιμωρ-ός, -όν, adj. with dat. *helping, to help.*
τιμωρῶ[έω], *help* (with dat.).
τις, τι, indef. pron. *some, someone, something;* neut. sing. as

adv. *in some respect, somewhat.*

τιτρώσκω, aor. *ἔτρωσα*, *damage, wound.*

τοι-όσδε, *-άδε*, *-όνδε*, *such, as follows.*

τοι-οῦτος, *-αύτη*, *-οῦτο*, *such.*

τοιουτότροπ-ος, *-ον*, *of this kind, in this way.*

τολμῶ[άω], *dare, submit to.*

τόξευμα, *-τος*, n. *arrow.*

τοξότ-ης, *-ου*, m. *archer.*

τόσ-ος, *-η*, *-ον*, *so great, so much.*

τοσ-οῦτος, *-αύτη*, *-οῦτο*, *so great;* in pl. *so many.*

τότε, adv. *then;* **ἐν τῷ τότε**, *at that time.*

τραυματίζω, *wound*

τραχ-ύς, *-εῖα*, *-ύ*, *rough, rugged.*

τρεῖς, *τριῶν*, *three.*

τρέπω, *τρέψω*, *ἔτρεψα*, pf. mid. *τέτραμμαι*, *turn, rout.*

τριάκοντα, *thirty.*

τριακόντορος (**ναῦς**), *-ου*, f. *thirty-oared ship.*

τριήραρχ-ος, *-ου*, n. *commander of a trireme.*

τριηραρχῶ[έω], *command a trireme.*

τριήρ-ης, *-ους*, f. *galley* (with three banks of oars), *trireme.*

τρίς, adv. *three times.*

τρίτ-ος, *-η*, *-ον*, *third.*

Τροιζ-ήν, *-ῆνος*, f. *Troezen.*

τροπαῖ-ον, *-ου*, n. *trophy.*

τρόπ-ος, *-ου*, m. *way, character.*

τροφ-ή, *-ῆς*, f. *food, sustenance.*

τυγχάνω, aor. *ἔτυχον*, with gen. *obtain;* with ptcp. *happen to ..., chance to....*

τυρ-ός, *-οῦ*, m. *cheese.*

Τυρσην-ός, *-ή*, *-όν*, *Tyrrhenian, Etruscan.*

τύχ-η, *-ης*, f. *chance, circumstances, fortune;* **κατὰ τύχην**, *by chance.*

ὑβρίζω, aor. *ὕβρισα*, *be,* or *grow, arrogant.*

ὑγι-ής, *-ές*, *healthy, honest.*

ὕδωρ, *ὕδατος*, n. *water.*

ὕλ-η, *-ης*, f. *wood, forest.*

ὑλώδ-ης, *-ες*, *woody.*

ὑμεῖς, *ὑμῶν*, see *σύ.*

ὑμέτερ-ος, *-α*, *-ον*, *your.*

ὑπάρχω, *be, be ready, be already, be to the advantage of* or *open to;* **τὰ ὑπάρχοντα**, *possessions, advantages.*

ὑπεκπέμπω, *send out secretly.*

ὑπέρ, prep., with accus. *beyond, more than;* with gen. *above, over.*

ὑπεραυχῶ[έω], *be arrogant.*

ὑπερφέρω, aor. ptcp. pass. *ὑπερενεχθείς*, *carry across.*

ὑπισχνοῦμαι[έομαι], aor. *ὑπεσχόμην*, *promise.*

ὑπό, prep., with accus. *under, towards, about;* with gen. *by, owing to;* with dat. *under.*

ὑποδε-ής, *-ές*, *inferior, weak.*

ὑποθορυβῶ[έω], *raise a clamour.*

ὑπόλοιπ-ος, *-ον*, *remaining, left.*

ὑπομένω, *ὑπομενῶ*, *await* (the attack of), *stand one's ground.*

ὑπόμνησ-ις, *-εως*, f. *reminder.*

ὑπονοῶ[έω], *suspect.*

ὑποστρέφω, *turn round.*

ὑπόσχεσ-ις, *-εως*, f. *promise.*

ὑποφεύγω, *shrink from.*
ὑποχωρῶ[έω], *retreat, give place.*
ὑποψί-α, -ας, f. *suspicion, ill-feeling.*
ὑσταραῖ-ος, -α, -ον, *on the next day*; **ἡ ὑστεραία**, *the next day.*
ὕστερ-ος, -α, -ον, *later*; adv. ὕστερον.
ὑφίστημι, 2nd aor. ὑπέστην, *set under*; in intr. tenses and mid. *undertake.*
ὕφυδρ-ος, -ον, *under water.*
ὕψ-ος, -ους, n. *height.*

φαίνω, aor. ἔφηνα, fut. pass. φανήσομαι, aor. pass. ἐφάνην, *show*; in mid. and pass. *show oneself, be shown, seem, appear.*
φανερ-ός, -ά, -όν, *clear, conspicuous.*
Φάρ-αξ, -ακος, m. *Pharax.*
φαῦλ-ος, -η, -ον, *bad, inferior.*
φείδομαι, *spare* (with gen.).
φέρω, οἴσω, ἤνεγκον or -α, *bear, carry*; **ῥαδίως φέρω**, *take lightly* or *calmly.*
φεύγω, φεύξομαι, ἔφυγον, *flee.*
φημί, φήσω, ἔφην, *say.*
φθάνω, aor. ἔφθην or ἔφθασα, *do first, be beforehand with, forestall*; in ptcp. *beforehand.*
φιλί-α, -ας, f. *friendship.*
φιλῶ[έω], *love, be wont, be used.*
φοβῶ[έω], *terrify*; in mid. and pass. *fear.*
φοιτῶ[άω], *go to and fro, come often.*
φράσσω, aor. ἔφαρξα, *block.*
φρούρι-ον, -ου, n. *fort, garrison.*
φρουρ-ίς (ναῦς), -ίδος, f. *guard-ship.*
φρουρ-ός, -οῦ, m. *guard*; in pl. *garrison.*
φρουρῶ[έω], *keep guard over, blockade.*
φυγ-άς, -άδος, m. *exile.*
φυγ-ή, -ῆς, f. *flight.*
φυλακ-ή, -ῆς, f. *guard, blockade, post.*
φυλακτήρι-ον, -ου, n. *outpost, fort.*
φύλ-αξ, -ακος, m. *guardian, guard.*
φυλάσσω, aor. ἐφύλαξα, *keep guard over, keep watch*; in mid. *beware of*, (with gen.) *be anxious about.*
φύσ-ις, -εως, f. *nature.*
φύω, 1st aor. ἔφυσα, 2nd aor. ἔφυν, *produce, grow*; pf. πέφυκα, *be by nature, be naturally inclined.*

χαλεπ-ός, -ή, -όν, *difficult, dangerous.*
χαλεπότ-ης, -ητος, f. *difficulty.*
Χαλκιδ-εῖς, -έων, m. pl. *Chalcidians* (in Chalcidice).
Χαλκιδικ-ός, -ή, -όν, *of the Chalcidians* (in Euboea).
χαρίζομαι, *grant a favour.*
χάρ-ις, -ιτος, f. *thanks, gratitude.*
Χάρυβδ-ις, -εως, f. *Charybdis.*
χειμ-ών, -ῶνος, m. *storm, winter, wintry weather.*
χείρ, χειρός, f. *hand*; **ἐς χεῖρας ἐλθεῖν**, *to come to close quarters*; **χεὶρ σιδηρᾶ**, *grappling-iron.*

χειροῦμαι[όομαι], *overpower, capture.*

χίλι-οι, -αι, -α, *a thousand.*

Χῖ-ος, -α, -ον, *of Chios.*

χλωρ-ός, -ά, -όν, *green.*

χοῖν-ιξ, -ικος, f. *quart.*

χρή, impf. χρῆν or ἐχρῆν, impers. *it is necessary.*

χρῆμα, -τος, n. *thing, matter;* in pl. often *money.*

χρόν-ος, -ου, m. *time.*

χρῶμαι[άομαι], inf. χρῆσθαι, *use* (with dat.).

χώρ-α, -ας, f. *country;* **κατὰ χώραν**, *in one's place.*

χωρί-ον, -ου, n. *place, ground.*

χωρῶ[έω], *go, advance.*

ψευδ-ής, -ές, *false, lying.*

ψηφίζομαι, *vote, vote for.*

ψιλ-ός, -ή, -όν, *light-armed.*

ὧδε, adv. *thus, so.*

ὠθῶ[έω], ὤσω, ἔωσα, act. or mid. *drive back.*

ὥρ-α, -ας, f. *season.*

ὡς, adv. *as;* with fut. ptcp. *intending to;* conj. *that, when;* with numbers, *about.*

ὥσπερ, conj. *just as.*

ὥστε, conj. *so that, on condition that*, with inf. or indic.

ὠφέλιμ-ος, -η, -ον, *useful, advantageous.*

ὠφελῶ[έω], *help, assist.*

Printed in the United States
By Bookmasters